Robert Brudenell Carter

Eyesight, Good and Bad

A Treatise on the Exercise and Preservation of Vision

Robert Brudenell Carter

Eyesight, Good and Bad
A Treatise on the Exercise and Preservation of Vision

ISBN/EAN: 9783337405427

Printed in Europe, USA, Canada, Australia, Japan

Cover: Foto ©berggeist007 / pixelio.de

More available books at **www.hansebooks.com**

A TREATISE ON
THE EXERCISE AND PRESERVATION OF VISION.

BY

ROBERT BRUDENELL CARTER, F.R.C.S.,

Late Hunterian Professor of Pathology and Surgery to the Royal College of Surgeons of England; Ophthalmic Surgeon to St. George's Hospital; Corresponding Member of the Royal Medico-Chirurgical Society of Edinburgh.

WITH NUMEROUS ILLUSTRATIONS.

London:

MACMILLAN AND CO.

1880.

PREFACE.

A LARGE portion of the time of every ophthalmic consultant is occupied, day after day, in repeating to successive patients precepts and injunctions which ought to be universally known and understood. The following pages contain an endeavour to make these precepts and injunctions, and the reasons for them, plainly intelligible to those who are most concerned in their observance.

It is proper to mention that some of the diagrams in the second and third chapters have been taken, with slight alterations, from Mackenzie's *Physiology of Vision;* and also that I have derived valuable suggestions from *Die Pflege der Augen* of Professor Arlt.

69, WIMPOLE STREET,
December, 1879.

CONTENTS.

CONTENTS.

EYESIGHT: GOOD AND BAD.

CHAPTER I.

THE STRUCTURE OF THE EYES AND OF THEIR APPENDAGES.

I DO not intend, in this chapter, to enter with any minuteness into matters of anatomical detail, but merely to write such a general description of the parts and the construction of the organs of vision as I think every educated person should possess. In giving opinions about patients, and still more frequently in giving evidence in courts of law, medical men are often placed in positions of difficulty by the want of knowledge of those whom they address; and it is not long since I myself was asked, by a learned judge, why it was that pictures of the interior of the eye, which were before him as matters of evidence, were painted of a red colour. The obvious answer, that they were so painted because the structures represented were themselves red, appeared to be the very last which his lordship was prepared to

B

receive; and it was impossible not to feel that his power of appreciating my testimony would have been increased, if he had known something more about the eyes than the fact that they are somehow useful in seeing. In speaking to patients, and more especially in giving opinions or advice which they have to communicate to others, similar difficulties are of frequent occurrence; and, now that the schoolmaster is abroad, and that the elements of physiology will before long be taught in Board Schools, it will not be seemly that those who receive what professes to be a superior education should be left entirely unacquainted with facts which will be made familiar to people of smaller opportunities and of inferior social station. The organs that are subservient to the bodily functions which we prize most highly, and which are of inestimable importance to our welfare, cannot be unworthy of some study at the hands of every person of even ordinary intelligence.

The function of each eye, taken singly, and as a condition antecedent to the act of vision, is to form upon the retina, or nervous membrane which lines the back of the organ, a sharply defined inverted image of any object which is looked at; and, if we take the almost transparent eye of a white rabbit, removed from a recently killed animal, we may see this picture shining through its outer coats, and therefore plainly visible. The mode in which the image is formed, optically speaking, will be considered in a subsequent chapter; and it is sufficient here to say that in principle the eye almost precisely resembles the common camera-obscura

of the photographer, and that the image produced upon
the retina is precisely analogous to that which is produced upon the sensitive plate, in order that it may be
fixed there by the chemical action of light. By means
of the optic nerve, the retina, which receives the image,
is in direct connection with the brain; and it is the
brain which interprets the visual appearances and com-
pletes the act of seeing. How this completion is
effected we do not at present know; but we do know
that it is dependent upon the sharpness and completeness of the retinal image ; and that, if this image is
blurred or imperfect, accurate seeing is impossible. We
shall have no concern in these pages with the element
of consciousness or of intelligence, which forms part of
the visual act ; and we have only to take note of the
share in vision which belongs to the eyes themselves,
that is to say, of the retinal images, and of the circumstances which contribute to or detract from their perfec-
tion. Among these, the first to be considered are the
formation, shape, and proportions of the eyeball.

The human eye, as shown in an enlarged diagrammatic
horizontal section in Fig. 1, is a nearly spherical body,
which, in the adult, measures a little less than an inch
in diameter. It consists chiefly of transparent fluids
contained in membranes, which are called the *tunics* of
the eyeball.

Of these tunics, which lie over one another concen-
trically, like the scales of an onion, the external is
divided into two parts, called respectively the sclerotic
and the cornea. The sclerotic (S, Fig. 1) covers the

posterior four-fifths, the cornea (C, Fig. 1) covers the front of the eyeball only. The front portion of the sclerotic is visible through the conjunctiva, or delicate

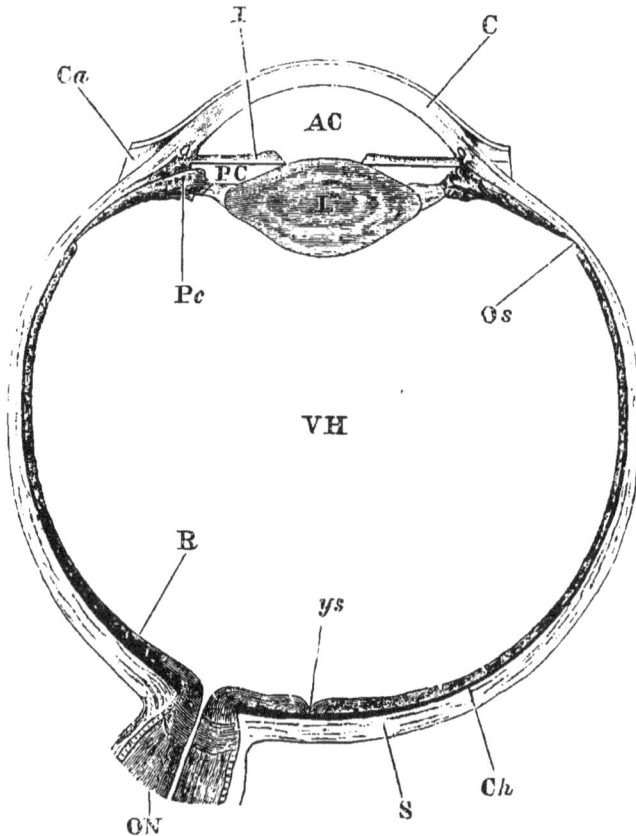

FIG. 1.

skin covering it, a portion of which is shown at Ca, and forms what is called the white of the eye. In early childhood it has a bluish tint, and towards the decline of life it generally becomes somewhat yellow.

The back of the sclerotic rests upon a cushion of fat which occupies the cavity of the orbit, and which surrounds and protects the optic nerve. In texture, the sclerotic is tough, flexible, and opaque. It is thickest at the hinder part of the eyeball, and becomes gradually thinner towards the front.

The cornea is inserted into the front margin of the sclerotic something in the manner of a watch-glass into its setting. It is not only as transparent as the most pellucid water, but as smooth as the most polished mirror; and to these qualities the eye is indebted for its brightness. The transparency, polish, and proper curvature of the cornea, are all of them of essential importance to vision.

These two membranes, taken together, complete the outer coverings of the eyeball, and they are maintained, by the fluids within their cavities, and by the blood which circulates in the vessels of the internal membranes presently to be described, in a certain state of tension or fulness, the degree of which can be estimated by pressing upon the eye with the points of the forefingers through the upper lid. By the maintenance of proper tension, the spherical shape of the eyeball is preserved; and this shape has led to a custom of describing the several parts of the organ by words borrowed from the language of geography. Thus, the eyeball as a whole is often called the "globe" of the eye; and its central points in front and behind are its anterior and posterior "poles." A line uniting these two poles is the axis of the globe; which, it may here

be observed, is not quite coincident with the axis of
vision hereafter to be mentioned.   Upon the axis of the
globe, a little behind its centre, and near the letters V H,
is a point which remains stationary in all movements of
the eye, and is called the centre of rotation.   An
imaginary plane, midway between the poles, and
dividing the globe into an anterior and a posterior
half, is called the equator ; and the halves themselves
are the anterior and posterior hemispheres.   It would
be possible, of course, to speak also of the upper and
lower hemispheres ; but all planes which are at right
angles to the equator, such, for example, as would
divide the cornea vertically or horizontally through its
centre, are called meridians.   For descriptive purposes,
too, the cornea is divided into imaginary quadrants;
and, by means of these several technicalities, it is possible to describe localities in the eyeball with great
precision and exactness.

   The inner surface of the sclerotic is lined by the next
tunic, C*h*, which is chiefly composed of a network of
blood-vessels, and is called the choroid.   The spaces
between the blood-vessels of the choroid are filled up
by a dark coloured substance called pigment, which
consists of fine granules, and which, together with the
circulating blood, gives to the interior of the eye, as
shown by the ophthalmoscope, a reddish brown or red
tint, which varies, in some degree, with the complexion
of the individual.   Somewhat in front of the equator
the choroid becomes thicker, and its inner surface is
thrown up into a circle of plaits or folds, P*c*, which are

known as the ciliary processes, and which constitute, collectively, the ciliary body. These processes, in the human eye, are seventy in number, and they form a ring or circle about a fifth of an inch in breadth, extending forwards from their origin to the insertion of the cornea. The pigmentation of the choroid renders it a perfectly opaque lining to the sclerotic, so that no light can enter the cavity of the eye excepting through the cornea.

The front border of the ciliary body is closely united with both the cornea and the sclerotic at their line of junction, and from this line the choroid is prolonged in the shape of a circular vertical curtain, I, called the iris, which has a central perforation, and which bears to the cornea something of the relation of the dial of a watch to the glass. The space between the iris and the cornea, A C, is the anterior chamber; and this, together with the smaller space or posterior chamber, P C, behind the iris, is filled by a clear fluid, called the aqueous humour, which consists almost entirely of pure water, and which is in contact with the iris on both sides. The iris is so called from the various tints, such as blue, grey, brown, hazel, and others, which it displays in different people; and, as seen through the cornea and the intervening aqueous humour, it forms the most conspicuous part of the eye. The central perforation in the iris is the pupil, a circular opening of variable size, and, in young and healthy eyes, of a bright black. The pupil undergoes dilatation in passing from a bright into a dull illumination, and in looking at distant objects;

and it contracts in passing from a lesser to a greater
brilliancy of light, or in looking at near objects, and
during sleep. The dilatation and contraction of the
pupil are performed by two sets of muscular fibres
which enter into the structure of the iris, the circular
and the radiating; and the changes which are thus
produced serve to regulate, and approximately to render
uniform, the quantity of light which is admitted into
the interior of the eye. Like the choroid, the iris is
abundantly supplied with pigment, especially on its
posterior surface; and it is thus rendered wholly impervious to light, which finds entrance into the eye
only through the pupil. The pupil appears black, because the cavity inclosed by the dark choroid and iris
contains transparent humours, which suffer light to pass
into them freely, but reflect back little in return, and
that little only in such a direction that it eludes observation unless looked for in a particular way. If we
look from a gas-lighted place at a hole opening into an
adjoining chamber, we receive but little light from
thence, even although the hole may allow it to enter
freely.

On the inner surface of the choroid, and closely in
contact with it, we find the retina, R, the third and
most important of the ocular tunics, to which, indeed,
the others are merely protective or containing membranes. The retina is the immediate continuation of
the optic nerve, ON, which extends from the brain to
the eyeball, perforates the sclerotic and choroid, and
immediately spreads out into a thin lamina over the

surface of the latter. The point of entrance of the optic nerve is nearly on the horizontal meridian of the eyeball, and about the tenth of an inch inwards from the posterior pole, so that it is the right eye which is represented in the figure. The office of the retina is to receive the pictures which are formed within the eye by the light reflected from objects of vision, and, through the medium of the optic nerve, to transmit the resulting visual impressions to the brain. Precisely as the sense of touch is not diffused uniformly over the surface of the body, but is more acute in some parts, as in the finger tips, than in others, as in the back of the hand ; so also the retina is not equally sensitive to luminous impressions over its whole surface, but in the highest degree in the vicinity of the posterior pole, in a part called the yellow spot, ys. A line drawn from any object looked at to the yellow spot is called the "axis of vision ;" and from the spot the sensitiveness of the retina gradually diminishes to the equator. The retina does not extend so far forwards as the choroid, but terminates a little in front of the equator, at the posterior border of the ciliary body, by an irregular margin, Os, the indentations of which have received the name of *Ora serrata.*

Immediately behind the iris, and in contact with the margin of the pupil when it is contracted, is the lens, or crystalline lens, L, a solid body which is inclosed in a delicate, transparent, and structureless membrane, the capsule, and is connected, through this, by an equally delicate membrane, the suspensory ligament, with the anterior border of the ciliary processes. In shape, the lens resembles an ordinary biconvex glass lens, except that it is less strongly curved in front than behind. In youth, it is a soft or moderately firm and highly elastic body, perfectly transparent and colourless, and as bright as the brightest crystal. With the advance of life it becomes harder, and sometimes of a slightly yellow tint, without losing its transparency ; but in old age it often becomes opaque or nearly so, a change which constitutes the affection known as senile cataract. The opaque lens varies in colour from a yellowish brown to a grey or white tint, and in some few cases has been almost black. Whatever its colour, it prevents vision by intercepting the passage of the rays of light to the retina; but the sight may in such cases be restored by removing the opaque lens from the eye by a surgical operation. Its place will then be supplied by watery fluid ; and, as this does not possess the refracting power of the organ which it replaces, vision is indistinct unless aided by a convex lens of the kind known as a cataract glass. Such a lens, when properly selected, performs, to some extent, the function of the natural lens which has been taken away.

The large cavity of the eyeball behind the lens, which is called the vitreous chamber, is filled by a colourless, transparent, gelatinous substance, V H, somewhat resembling the white of egg, but less fluid, and known as the vitreous humour. Its office appears to be to maintain, between the crystalline lens and the retina, an interval, like that which, in a camera, intervenes between the lens which forms the image and the screen upon which it is received. This interval, in the camera, might be filled with fluid as well as with air, except that the substitution would render it necessary to place the screen farther from the lens.

The vitreous humour, although to the naked eye it appears homogeneous and transparent, yet contains numerous fine filaments distributed throughout its substance; and these filaments carry upon them the remains of the cells from which they were produced, or the germs of other cells which might in favourable circumstances be called into activity. The cells and filaments are themselves transparent, but it sometimes happens that the degree or index of their refraction differs from that of the fluid in which they float, and they are then liable to cast shadows upon the retina, precisely as a glass rod might cast a shadow when immersed in a tumbler of water. Such shadows are projected outwards by the sense perceptions as visible floating objects. The magnitude of each shadow will depend partly upon the magnitude of the cell or filament producing it, and partly upon the distance of the latter from the retina, or screen which receives the shadow; and hence such shadows are largest and most conspicuous in short-sighted eyes, which, as a matter of formation, are longer from front to back than others. The shadows appear as floating fibres or particles, which are most plainly seen against a white cloud or white wall, or other bright and uniform surface. They have received the name of *Muscæ volitantes*, and it is one of their characteristics that they never actually shut out any point which is being looked at, but float a little above, below, or to one side of it ; the reason of this being that the filaments do not appear to exist in the axis of vision, but only in the lateral parts of the humour. Muscæ may be rendered more than usually conspicuous by any circumstances which alter the density of the fluid part of the humour, and thus increase, either temporarily or permanently, the difference of refraction between this fluid part and the filaments or cells. Muscæ exist, more or less, in all eyes, and may easily be found by looking at a white cloud through a pinhole in a card; but they are usually harmless, and only require mention on account of the anxiety which they sometimes cause to persons who are unacquainted with their nature.

The free movements of the eyes are of such a kind that they involve no change of place, but only rotation upon a central point in each eyeball. The pupils can be directed upwards, downwards, outwards, and inwards, and also in intermediate directions, as upwards and outwards, without the eyeball as a whole changing its position. These movements are accomplished by the agency of six muscles. By muscles, we mean certain flat or round structures, composed of fibres or threads, which, under the influence of the will or of other stimuli, can be made to contract or shorten themselves, and which relax and return to their former length as soon as the will or the stimulus ceases to act. By such shortening or contraction, all our voluntary movements

are performed; and there are also muscles of a somewhat different kind, which act independently of the will.
The flesh of animals, which we use as food, consists
almost entirely of muscles, and these may easily be
unravelled into the fibres of which they are composed.
Four of the six muscles of each eye are called the *recti*,
or straight muscles; and these have their fixed points
of insertion behind the eyeball, upon the bones of the
skull at the apex of the orbital cavity. Passing forwards, they embrace the eye closely; and are inserted
into the sclerotic, one on the inner, or nasal side, one on
the outer, or temporal side, one above, and one below,
at a distance of about a quarter of an inch from the
margin of the cornea. When the inner straight muscle
contracts, it rotates the anterior pole of the eye inwards,
or towards the nose; and the posterior pole outwards,
or towards the temple. When the contraction ceases,
the eye is brought back, by the external straight muscle,
to its position of rest in the centre of the lid-opening.
The other two muscles, which from their respective
directions are called the superior and inferior oblique,
have their fixed points in front of the equator of the
eye, on the inner side of the orbit, the one above, the
other below, the inner angle of the lids; and they are
inserted into the sclerotic, on the outer side, between
the equator and the posterior pole. By the single or
combined contraction of one or more of these pairs of
muscles, the eyeballs can be moved in every direction
with greater rapidity than any other portion of the
body; and there is no other group of muscles which, in

the waking condition, is called upon to perform so many and such fine gradations of movement as those of the eyes. The stimulus by which these movements are called for is mainly that of the images produced upon the retina, and communicated as visual sensations to the consciousness; and the response to such sensations is often almost inconceivably rapid. For example, a case is recorded in which a man who was seated on a chair was stabbed with a penknife by another who was standing over him, and who struck downwards upon his eye. The blade pierced the upper lid, but pierced the eyeball below the cornea; so rapid had been the action of the superior straight muscle in rolling the globe upwards in order that it might be less exposed to injury. Nearly all other voluntary muscles require to be rested, even during waking hours, by occasional repose or by change of effort; but from this requirement those which move the eye appear to be practically exempt.

Next after the muscles, the parts most important from our present point of view are the eyelids. Each of these is formed or moulded upon a thin plate of substance resembling cartilage, or gristle, which preserves the shape and firmness of the lid. They are covered externally by very fine and more or less wrinkled skin, internally by a delicate membrane called the conjunctiva. Between the skin and the cartilage the lids are crossed by thin bundles of muscular fibres, directed in curves from the inner to the outer angle, and extending from near the margin of the lids to that of the orbital opening both above and below. These fibres collectively form the muscle which closes the lids, and which is named the *orbicularis* by anatomists. The upper lid is raised or held open by another muscle, the antagonist of the orbicularis, which passes along the roof of the orbit from the bones of the skull, and is inserted into the whole length of the upper border of the cartilage ; but the lower lid sinks to the open position by its own weight, as soon as the action of the closing muscle ceases. This muscle is not only called into action during sleep, or to shut the eyes designedly, but also for the performance of that very frequent involuntary closure and re-opening of the lids by which the surface of the eyeball is moistened, and freed from particles of atmospheric dust or other impurity ; a provision which is of the highest importance for the preservation of the polish and transparency of the cornea. From the outer borders of the lids spring the eyelashes, which afford protection against the entrance of coarse dust, insects, and other foreign bodies or intruders ; and, immediately within the eyelashes, on the very margins of the lids, are the openings of numerous fat glands which are situated in the substance of the cartilage. These glands, which are called *Meibomean*, after the name of their discoverer, furnish a greasy secretion which lubricates the edges of the lids, and which acts as a barrier to the escape of moisture from the eyes over the cheek.

The free mobility of the lids over the eyes is permitted by the conjunctiva, which, after lining the lids as described, is formed into loose folds at their attached margins, and from these is reflected, or carried over, to the surface of the eyeball itself, where it covers the whole of the front portion of the sclerotic, and, as shown at C *a*, furnishes an extremely delicate coating even to the cornea. The conjunctiva secretes a portion of the fluid by which the surface of the eyeball is moistened; but the greater part of this fluid is furnished by the lacrymal or tear gland, which is situated above and to the outer side of the globe in the cavity of the orbit, and discharges its secretion by several fine tubes beneath the upper and outer portion of the upper lid.

The tear secretion, when formed only in natural quantity, is partly evaporated by the air, and partly absorbed by the conjunctival surface ; but the residue is conveyed away from the eye by a series of tubes which conduct it into the nose, so that the tears do not flow over the cheeks except during weeping or other periods of excessive secretion. The tear passages commence near the inner angle of the lids by two fine round white openings, which are plainly visible when either the upper or lower lid is drawn somewhat outwards, and which lead into two delicate tubes or channels. These channels terminate in the tear bag, a small pouch or cavity beneath the skin of the inner side of the nose, immediately below the inner angle of the eyelids. From this pouch a single duct leads downwards into the nose. The tear passages are liable to become closed or obstructed by local inflammation ; and, when this happens, a troublesome overflow of moisture is produced. If this condition is not rectified, it may be followed by the establishment of a new and often permanent opening upon the cheek, which becomes a source of much disfigurement and inconvenience.

CHAPTER II.

IT has been stated in the foregoing chapter that vision is dependent upon the formation of images upon the retina, by light reflected from the objects which are seen ; and, before considering how these images are produced within the eye, it is necessary to explain the formation of analogous images by artificial means.

We mean by light a mode of force which is resolvable into molecular wave movements, and which is well known to us in many of its manifestations. Disregarding the wave movements, an account of which is not necessary for the purpose now before us, we may speak of light as a force which emanates from luminous bodies, and which travels in straight lines in all directions. If we light a candle, and place it in the middle of an otherwise dark room, all parts of the room equidistant from the flame will be equally illuminated ; and the amount of light falling upon equal surfaces will be inversely as the squares of their distances from the source of illumination.

If we take two equal sheets of paper, say each six inches square; and place one of them six inches from the candle and the other twelve inches from it, the latter will only receive one fourth as much light as the former; and, if we remove the latter to a distance of eighteen inches, it will then only receive one ninth part as much light as the former. For equal distances, however, the illumination afforded by a central source of light is equal in all directions.

In considering the course of the rays of light which proceed from any luminous object of appreciable magnitude, such as a candle flame, we must not regard this as a single source of radiation, but as an assemblage of luminous points of infinite minuteness, from every one of which rays are given out equally in all directions. The rays from these several points must necessarily cross one another as they diverge.

As long as the rays of light pass only through the atmosphere, they continue to travel in straight lines; but, when they fall upon any other substance, they may be variously diverted from their original course. If the substance is opaque, the rays falling upon it are arrested, and are partly absorbed into its material, partly reflected or turned back; and the relative proportion in which one or the other of these results is produced is greatly dependent upon the quality of the surface, whether it be dull or polished. If the substance is transparent, much of the light passes through it, either with or without a change in its direction, but part is always reflected.. It is by means of the light reflected

from surfaces, whether they are opaque or transparent,
that images are formed upon the retina, so that the
surfaces themselves are rendered visible. A substance
capable of absorbing or transmitting all the light which
it received, if any such substance existed, would be
invisible.

A ray of light which strikes upon any surface, and is
turned back, obeys precisely the same physical law as a

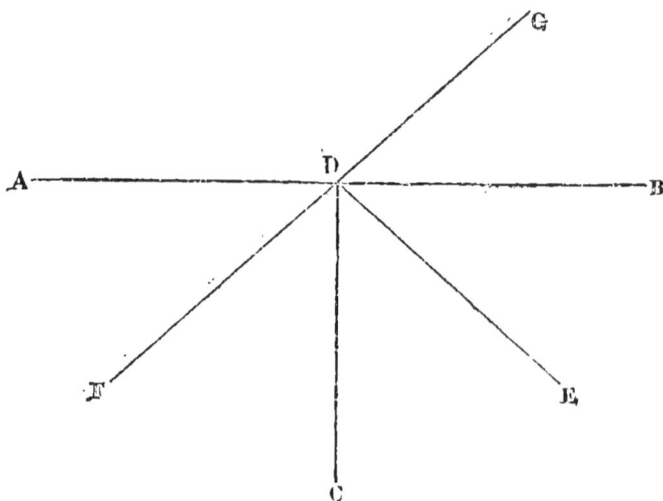

FIG. 2.

billiard ball under the same conditions ; that is to say,
the angle of reflection is equal to the angle of incidence.
In Fig. 2 the line A B represents a reflecting surface, C D
is a line perpendicular to this surface, and E is a luminous
point. A ray from the point E, which falls upon the
surface A B in the point D, making with the surface the
angle E D B, and with the perpendicular the angle E D C,

will be reflected in the direction D F, making the angle
A D F equal to B D E, and the angle C D F equal to C D E.
When the reflection takes place from a polished surface,
and is tolerably complete, the reflected ray appears to
an observer who is situated in its course, as at the point
F, to proceed from a point in a backward prolongation
of the line D F, and as much behind the surface A B as
the real source of light is in front of it. Thus, if A B
were a mirror, and E a candle, an observer at F who was
unconscious of the presence of the mirror, and from
whom the actual candle was screened, would think that
it was standing at the point G, instead of in its actual
situation. Illustrations of the operation of this law
must be familiar to every one ; and it forms the basis
of a great number of common optical illusions,
employed by conjurors and others.

A ray of light which falls upon a transparent substance, and which passes through it, or is said to be
transmitted, is seldom transmitted in an unchanged
direction. This can only happen, indeed, under one of
two conditions: first, if the new medium is of the same
refracting power as the atmosphere; secondly, if the
ray falls perpendicularly upon the surface. Under all
other conditions, the transmitted ray is more or less
bent out of its course; and it is then said to be refracted.
The law is that light, passing in an oblique direction,
either out of a rare into a dense medium, or out of a
dense into a rare medium, is refracted in different
degrees according to the relative refractive powers of
the two media ; towards the perpendicular, if the new

medium is the more dense, and from the perpendicular, if the new medium is the more rare.

Let A B, Fig. 3, represent a ray of light, passing

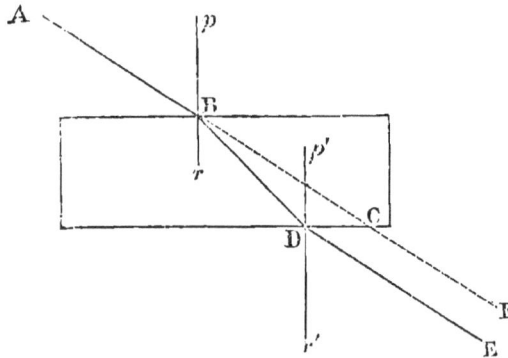

FIG. 3.

through air, and incident obliquely on the surface of water at B. Instead of pursuing its original course to C,

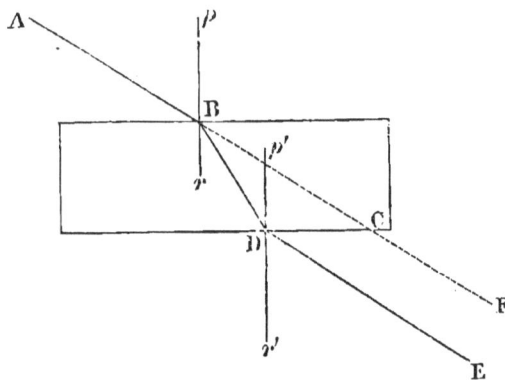

FIG. 4.

it will be refracted into the direction B D, and drawn towards $p\,r$, a line perpendicular to the surface of the water at the point of incidence B. Let A B, Fig. 4, represent a ray of light falling in like manner obliquely at B on the surface of a denser medium than water, say glass. It will be drawn still more towards the perpendicular $p\, r$, being refracted into the direction B D, instead of pursuing its original course to C. If the dense medium is bounded by plane surfaces, parallel to each other, as is represented in Figs. 3 and 4, on quitting the dense medium, to enter one which is less dense, say air, the ray will undergo a second and opposite refraction. The ray B D, quitting obliquely the second surface of the dense medium, is refracted from the perpendicular $p'\, r'$, and takes the direction D E, which, in both cases, is parallel to C F, the original course of the ray. By comparing the angle of incidence B D p' with the angle of refraction E D r', in the two cases, it will be evident that the refraction of the ray B D, on quitting the second surface, is greater when the refracting medium is glass than when it is water.

It being understood that different transparent bodies, whether solids or liquids, possess very different degrees of refractive power, not measurable by mere density, and that the amount of the deviation of the refracted ray from its original course is always proportionate to the refractive power of the medium, it is next necessary to explain that by varying the obliquity of the surface of the refracting medium, in respect to the incident ray or rays, we are able to produce any particular deviation we wish to obtain, whether in respect of degree or direction. If we wish to produce a great degree of deviation, we give an increased obliquity to the refracting surface ; if

we wish a small degree of deviation, we employ a refracting surface of which the obliquity is slight. This
may be illustrated by Fig. 5, in which A, B, C, D, are
supposed to be rays of light passing from a rare into a
dense medium. The ray A meets the surface of the
medium perpendicularly, that is, without any obliquity;
therefore, there is no deviation. The ray B meets the

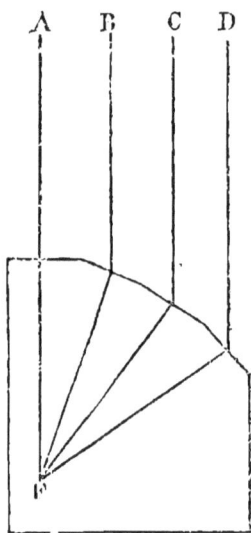

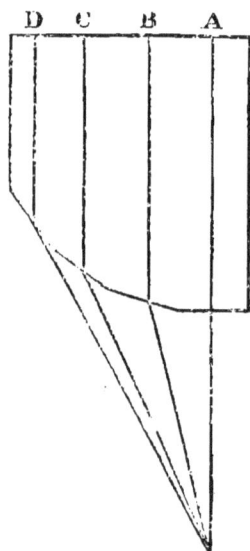

FIG. 5.   FIG. 6.

refracting surface with a slight obliquity; therefore, there
is a small degree of deviation. The deviation of C from
its original direction is greater than that of B, and that
of D greater than that of C, in proportion to the increasing obliquity of the refracting surface. In the figure, all
the rays are represented as coming to a focus F, but this
is not essential to the principle.

If the rays are passing from a dense into a rare medium, as in Fig. 6, the same principle is applicable. To produce a great degree of deviation, we must give an increased obliquity to the surface relatively to the ray. In A the deviation is null, because there is no obliquity. In B, C, and D, the deviation increases with the obliquity of the refracting surface.

A slight consideration of the facts above stated will

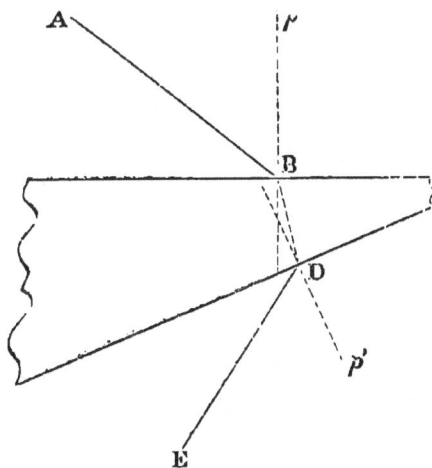

FIG. 7.

be sufficient to suggest that, if the surfaces of a dense medium be inclined to one another, the refraction which the ray will undergo at the second surface, instead of restoring it to its original course, as was the case in Figs. 3 and 4, where the surfaces were parallel, will augment its deviation. Fig. 7 represents a dense medium, with two surfaces inclined to each other. The ray A B is at the first surface refracted into the direction B D, towards the perpendicular p ; but at the second surface, it is carried into the direction D E, away from the perpendicular p'. The second refraction, then, carries the ray still farther from its original course. By a simple application of this principle, it is manifest that, when a pencil of rays passes through a refracting surface or surfaces, the rays may be bent into determinate directions by determinate outlines of the surfaces, and that variations in the outline will produce corresponding variations in the refraction. It is also manifest that the surface of any medium intended to bring a pencil of rays to a focus must be a curve ; and, further, that the surface of a dense medium employed for this purpose must be convex, both in the case in which we wish to produce convergence by transmitting the rays from a rare into a dense medium, and also when the transmission is from a dense into a rare medium. We may now apply the principles above explained to the determination of the figure of a dense medium, which shall be fitted for collecting rays to a focus.

Let the luminous object be very remote, so that the rays proceeding from it may be considered as parallel ; for, although all the unrefracted rays existing in nature are more or less divergent, yet, when they proceed from great distances, their actual deviation from parallelism is insensible. Let A, B, C, D, E, Fig. 8, represent these rays. Only one of them, C, by continuing its straight course, can arrive at the point F. The surface of the dense medium should be presented at right angles to this ray, at i, so that the ray may pass through the medium without

deviation. Those rays, B and D, which are situated near
to the direct or central ray C, will require but a small
degree of refraction to enable them to reach the focus F,
and this small refraction will be produced by a small
degree of obliquity in the dense medium at the points
h and *k*. In proportion as the rays A and E are more
distant from the central ray, a greater amount of refraction, and consequently a greater obliquity of surfaces at
g and *l* will be required to bring them to the same focus.
On the presumption that the rays passed through a

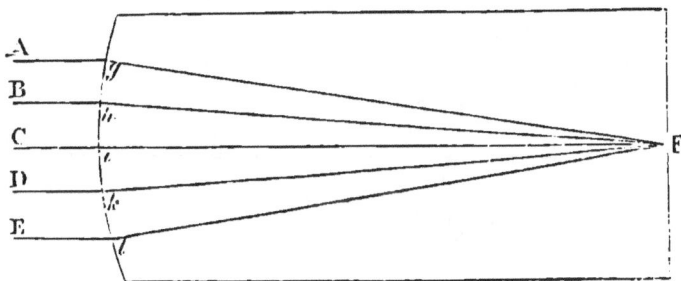

FIG. 8.

medium of uniform density, they would converge to a
focus, then, at F.

Let Fig. 9 represent the same rays A, B, C, D, E, entering
the same medium as before, but instead of the medium
being continued, let it be supposed to terminate at the
curved surface *m, n, o, p, q*, so that it now forms a double
convex lens. The central ray C proceeds at right angles
through both surfaces, and reaches F' or F, without
deviation. The rays B, D, A, E are refracted towards the
perpendiculars on passing into the dense medium at the

points *h, k, g, l,* but on quitting it they are refracted from the perpendiculars to the surface of the rare medium, at the points *n, p, m, q.* This new refraction increases the convergence of the rays, and brings them to a focus F′, nearer to the dense medium than the former focus F.

The result of the continual change of direction in the refracting medium is a regular curvilineal surface, approaching to the spherical. By giving such surfaces to refracting substances they become fitted to produce, with

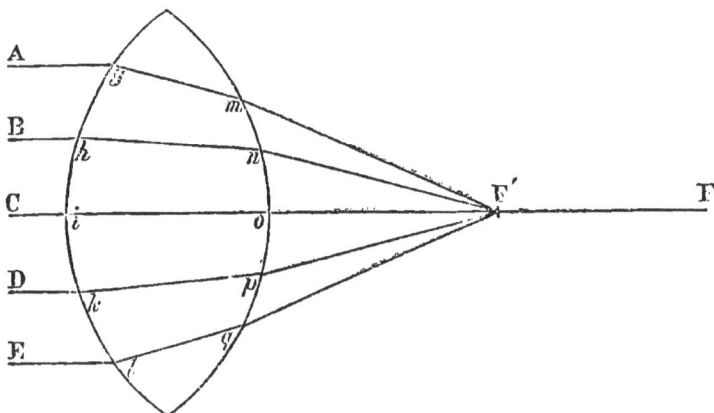

FIG. 9.

more or less exactness, the convergence of parallel rays to a focus ; and, by making the dense medium convex on both sides, both conspire to produce the desired effect.

The distance of the focus behind the medium depends on the refracting power of the substance employed, and on the degree of convexity of its surfaces. The greater the convexity of the two surfaces, and the greater the refractive power, the nearer will be the focus.

We have next to apply the foregoing principles to the

formation of images by lenses ; and for this purpose it
is necessary first to consider the results of the ordinary transmission of light in straight lines from its
source. If we take a card, as A, Fig. 10, perforate it
by a central hole about one-tenth of an inch in diameter,
and then place it between a lighted candle and a white
screen, in an otherwise darkened room, we shall see an
inverted image of the candle flame upon the screen.
The light, radiating from the flame in every direction,
strikes partly upon the imperforated part of the card,

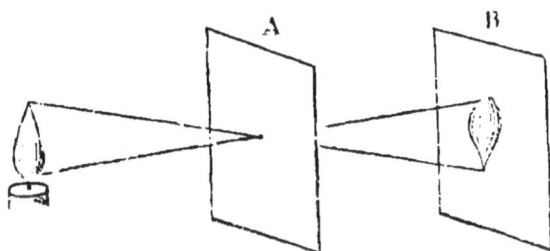

FIG. 10.

and is thereby either reflected or absorbed ; but that
which passes through the hole continues its original
direction unchanged. Hence, if we suppose the hole
to be opposite the centre of the flame, the ray of light
from the top of the flame, which descends to reach the
hole, continues descending after having passed through
it, and thus arrives at the lower part of the screen B ;
while that from the lower part of the flame, ascending
to reach the hole, arrives at the upper part of the screen.
The rays from intermediate points of the flame cross in
the same manner, but of course, strike the screen nearer

to its centre; and the rays from lateral points cross right and left in the same way. The result is that the surface of the screen is illuminated in a patch which copies the shape of the flame from which the rays proceed, but in an inverted position. The inversion may be still farther illustrated by taking another screen, and gradually interposing it between the flame and the perforation. If we bring this second screen slowly downwards, as it comes in front of the top of the flame, the bottom of the image will disappear. If we bring it upwards, the top of the image will disappear; and, if we bring it across from right or left, the opposite side of the image will in each case be obliterated. Hence we see that in every direction, laterally as well as vertically, the inversion of the image is complete. If we place two lighted candles before the card, two images will be formed upon the screen; and, if we move the candle which stands nearer to us, the position of the image which is farther from us will be altered, showing that the images are completely reversed. If we use a row of candles, rays of light flowing through the hole from all of them will form on the screen as many images as there are flames, each image being as clear and distinct as if there were only one. By this experiment we obtain evidence of the great tenuity of light, which allows the rays from several distinct sources to pass together through the hole, travelling on slightly different lines, and without any commingling or confusion.

The size of the image formed in the manner above described is dependent upon two factors; the distance

of the screen from the hole, and the distance of the
flame from the hole.   A glance at the figure will show
that, as the lines of light are continually divergent after
passing through the hole, they become more widely
separated the farther they go, and the size of the image
which they form must be increased in a corresponding
degree.   On the other hand, the nearer the flame itself
is to the hole, the greater will be the inclination of the
lines which proceed from its boundaries, and the more
rapid their divergence afterwards.   The law is, that the
linear magnitude of the image will bear the same proportion to that of the object, as the distance between
the aperture for transmission and the screen on which
the image is received bears to the distance between the
same aperture and the object.   The *absolute* magnitude
of the image, or the surface covered by it, increases
directly as the squares of the distances of the screen
from the aperture of transmission ; and decreases inversely as the squares of the distances of the object
from the same aperture.   Thus, at one inch from the
aperture, the image covers a certain extent of surface ;
at the distance of two inches it covers four times that
surface, at the distance of three inches, nine times that
surface, and so on.   If the object is at the distance of
one inch from the aperture, the image will have a certain
absolute magnitude ; if the object be removed to a distance of two inches, the image will be diminished to
one-fourth ; if the object be removed to the distance
of three inches, the image will be diminished to one-
ninth, and so on.

The images formed in this manner, or, as it is called, by radiation, are never of any great distinctness. It will be obvious that all their brightness is dependent upon the amount of light which passes through the small aperture described ; and it is easy to see that by diminishing this aperture we may diminish light to the extent of obtaining no image at all ; while, if we seek more light by enlarging the aperture, we render the outlines

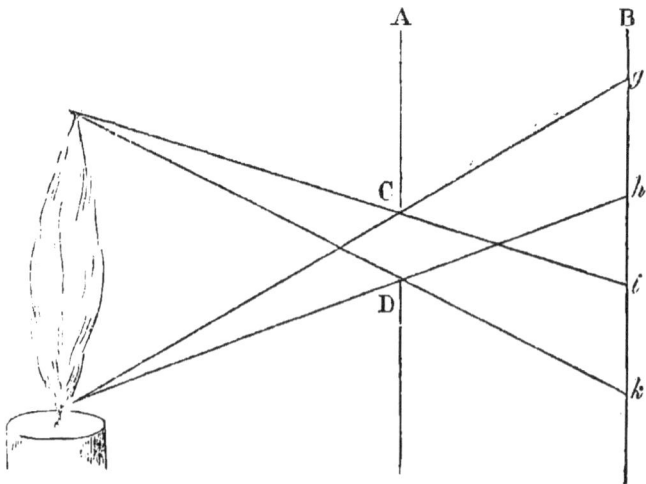

FIG. 11.

so indistinct that they soon cease to be recognisable. In Fig. 11 we have the same conditions as in Fig. 10, but with a larger aperture in the screen A. This aperture, C D, would then receive from every point of the flame not a ray of light only, but a cone of rays, and this cone would continue to widen its base as it proceeded towards the screen B, on which it would ultimately form a round patch of light, as at *g h*, *i k*, and

each of these patches would mingle with the similar
patches coming from other points of the flame, in such
a manner as to produce merely a general illumination
of the surface, without any definiteness of outline. In
order to obtain an image which is at once sufficiently
luminous and sufficiently defined in shape, it is necessary
to have recourse to some contrivance by which the rays
of light proceeding from each point will be united in a

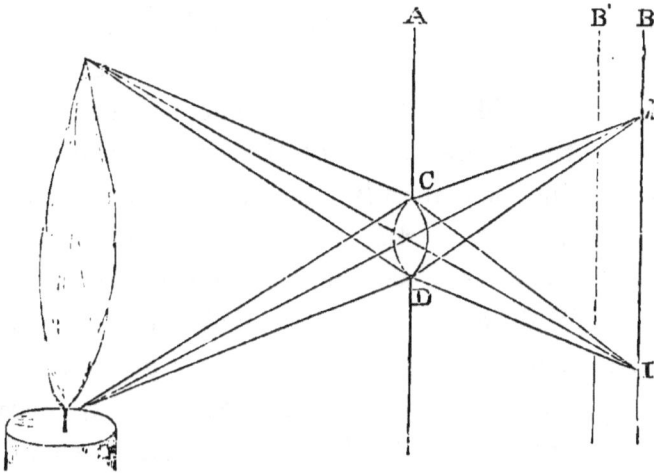

FIG. 12.

corresponding point upon the screen, and this requirement is fulfilled by means of that·bending or refraction
of the rays which we have seen to be a property of
lenses. Let us repeat, in Fig. 12, the conditions shown
in Fig. 11, with the exception that we fill the aperture
C D by a double convex lens, and that we place the
screen B at the focal distance of this lens. In these
circumstances, the cone of light issuing from the summit

D

of the flame, being refracted by the lens, will be reunited at the point I, the direction of the axial ray of
the cone remaing unchanged, but all the others being
brought to unite with it. The cone of light from the
base of the flame will in like manner be brought to
a focus at k, and the cones from every intermediate
point will form corresponding focal points, so that an
inverted image of the flame will appear upon the screen
B at Ik, the whole of the light from the flame which is
received by the surface of the lens will be concentrated
upon the space Ik, and the image will of course be
much more vivid than when formed by the mere exclusion of the lateral parts of the cones, as in Fig. 10.
On the principles already laid down, the shorter the
focal length of the lens, and therefore the nearer the
image, the smaller it will be. A powerful lens, or one
of short focal length, gives a small and very bright
inverted image; a lens of low power gives an image
which is larger, but which is less bright, because,
assuming the lenses to be of the same superficial
magnitude, each image is formed by the same quantity
of light.

A perfect illustration of the mode in which an image
is formed by a lens is furnished by the photographer's
camera, now so universally familiar. This instrument consists essentially of the lens and screen shown in Fig. 12,
with the difference that these form portions of a box
or chamber, that the screen is of ground glass, so that,
being transparent, the image can be seen through it,
and that the lens is furnished with an adjusting screw,

by which it may be *focussed*, as it is said, for objects at different distances. By means of such an apparatus, any one who so desires may become well acquainted, after a few simple experiments, with all the conditions under which inverted images are ordinarily produced.

Among the first things which may be observed by the aid of a camera is, that the nearer the object is to the lens, the greater must be the distance between the lens and the screen, in order that a clear and well-defined image may fall upon the latter ; and the reason of this is not far to seek. There is, for every lens, a constant distance at which it will bring to a focus rays which fall upon it in a state of parallelism. Let us suppose, in the case of a given lens, that this distance is ten inches. It is obvious that, if the rays which fall upon it are not parallel, but divergent or spreading out from their point of issue, a certain portion of the power of the lens will be consumed, so to speak, in rendering them parallel, before it can begin to render them divergent ; and hence their union in a focal point will be delayed, or will only occur farther away from the lens, than if they were parallel originally. In like manner, if the rays are already convergent when they reach the lens, part of its work will be already done ; and the focal union will occur sooner, or nearer to the lens, than if the rays were parallel. In estimating the power of a lens, we always take its focal length for parallel rays as the basis of computation ; and this is called its principal focal length, or, more commonly, its focal length only. It is, of course, invariable ; while the distances of its foci for convergent or divergent rays will depend, in every case, upon the degree of the deviation of these rays from parallelism. Strictly speaking, all light exists in nature in the form of divergent rays, but those which proceed from a far distant point to fall upon so limited an area as that of a small lens may, as already said, be considered and treated as parallel. As soon as the luminous point or other object approaches the lens, however, the divergence of the rays becomes very appreciable ; and so the camera, when arranged to give a clear image of the horizon, would give only a blurred and confused image of objects on the other side of a room. In order to render the latter image as clear as the former, either the distance between the lens and the screen must be increased, or else the power of the lens itself must be increased, as by the addition of a second one. Unless one or other of these changes were made, the screen would intercept the rays of light before they were brought to union, as if it were in the position shown by the dotted line B' in Fig. 12, and an imperfect or indistinct picture would be produced. We shall find, hereafter, that a similar provision for adjustment is necessary in the eye itself, in order that it may receive, with equal clearness, images of the many and variously distant objects to which it is from time to time directed.

CHAPTER III.

THE FORMS AND PROPERTIES OF LENSES.

WE have seen already, in Fig. 7, that a dense transparent medium, the two sides of which are inclined
towards one another, refracts a ray of light away from
its thinner and towards its thicker portion ; and to this
elementary fact all the varieties of refraction by lenses
may be reduced. The word lens is used to signify any
piece of glass or other transparent substance, which is
used for the purpose of producing refraction ; and it is
manifest that lenses may be of various figures. The
lens with inclined sides, which forms the basis of them
all, is called a prism, A, Fig. 13, and the other best
known forms are the following :—

A *Double-convex* Lens, B, is bounded by two convex
spherical surfaces, whose centres are on opposite sides
of the lens. It is equally convex when the radii of
both surfaces are equal, and unequally convex when
the radii are unequal.

A *Plano-convex* Lens, C, is bounded by a plane surface on one side, and a convex on the other.

A *Meniscus* (that is, a little moon, or crescent), D, is bounded by a concave and a convex surface, and these two surfaces meet, if continued, so that the element of convexity preponderates.

A *Double-concave* Lens, E, is bounded by two concave spherical surfaces, whose centres are on opposite sides of the lens.

A *Plano-concave* Lens, F, is bounded by a plane surface on one side, and a concave on the other.

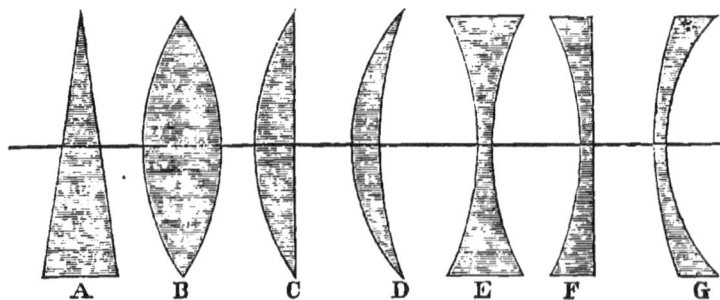

FIG. 13.

A *Concavo-convex* Lens, G, is bounded by a concave and a convex surface, the radius of the concave surface being shorter than that of the convex, so that the surfaces do not meet if continued, and the element of concavity preponderates.

It has already been explained that, when a ray of light passes through a prism, denser than the surrounding medium, the total deviation of the ray is in all cases from the vertex. The general effect of any lens may be understood by resolving it into two prisms. If the bases of the prisms, of which the lens is supposed to be

formed, be turned towards each other, the lens must
be convex, and the total deviation of the rays which

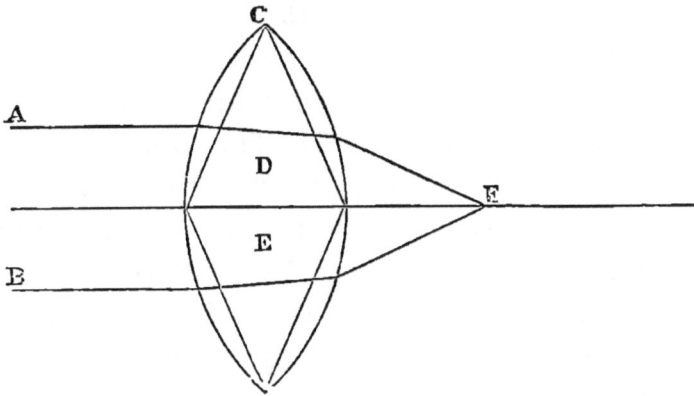

FIG. 14.

pass through it will be towards its central axis ; but if
the bases are turned from each other, the lens must be

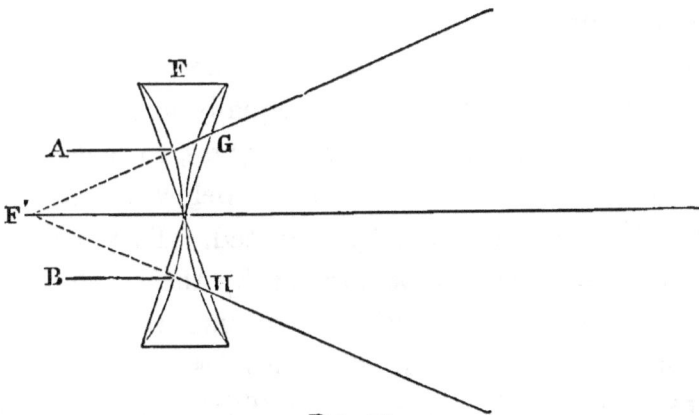

FIG. 15.

concave, and the rays of light will be bent from its axis.
The rays of light, A B, Fig. 14, are refracted by the

convex lens, C, precisely as they would have been by the two inscribed prisms, D, E ; and in the same way the concave lens, F, Fig. 15, resembles in its operation the effect of the prisms G and H.

On this principle, the six lenses, B, C, D, and E, F, G, Fig. 13, form two classes ; the first three being *convergent* and the last three *divergent.* The first three either cause rays of light to converge, or lessen their divergence; and the last three either cause them to diverge, or lessen their convergence. The lenses which are thinner at the edge than in the middle are convergent, and those which are thicker at the edge than in the middle are divergent. The first class are often called *magnifying glasses*, and are used by those whose eyes are too flat in shape, as well as by persons advanced in life ; while the second class are often called *diminishing glasses*, and are used to aid the distant vision of people who are short-sighted.

In all the foregoing varieties, the lenses are portions of spherical surfaces, and curve equally in all directions from their centres or axes to their edges, so that all the light which falls upon them, above, below, or on either side of the centre, is equally refracted. They are solid bodies, such as would be formed by the revolution of the sections, shown in Fig. 13, around the horizontal axis on which they are drawn. Besides these spherical lenses, much use has been made of late years of lenses of another form, which are portions of cylindrical surfaces, and which refract only those rays of light which fall upon them in certain definite directions. The

cylindrical lenses in common use are the plano-convex
and the plano-concave varieties; and these would be
formed, not by the revolution of sections C and F, Fig. 13,
but by building up these sections so as to give them an
appreciable height from the surface on which they rest.
In Fig. 16, A A represents a plano-convex cylindrical
lens, and the dotted lines show the cylinder of which it
forms a part. A plano-concave cylindrical lens is of

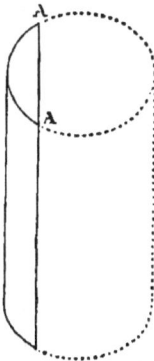

FIG. 16.

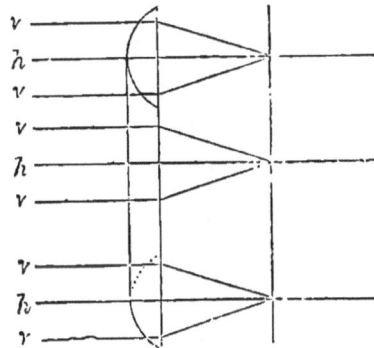

FIG. 17.

analogous form, only its curved or refracting surface
is concave instead of being convex.

The action of a plano-convex cylindrical lens upon
light is shown in Fig. 17. In this figure, the parallel
rays of light, *h h h*, which fall upon the convex surface of
the lens in a vertical plane, or in a plane coincident with
the axis of the cylinder of which it forms part, are not
refracted at all. They only encounter and pass through
a piece of glass with parallel sides, and this does not
appreciably change their direction. They enter the lens

parallel, and they emerge parallel on the other side.
But the rays vv, vv, vv, which fall upon the glass in
successive horizontal planes, that is, in planes perpendicular to the axis of the cylinder, encounter a strongly
curved refracting surface, and are brought to foci accordingly.   Hence, while a spherical convex lens, such as a
common magnifying glass, brings the rays which fall
upon it, say the rays of the sun when the lens is used
as a burning glass, to a focal *point*, a cylindrical convex
glass brings such rays only to a focal *line ;* since those
only are united which pass through it in a direction
transverse to its axis, and those which pass through it
in the same direction as its axis are left without any
deviation from their former course.

We have already seen that the power of a convergent
or convex lens is expressed by the distance of its principal focus, that is, of the point at which it would cause
parallel rays to unite after they had passed through it,
as at F, Fig. 14.   The power of a divergent lens is
expressed, in a similar manner, by the distance of the
point F', Fig. 15, from which parallel rays appear to
diverge after they have passed through it. In both cases
this distance is called the focal length; and a convex
and a concave lens of similar focal length, if combined,
completely neutralise one another, and produce the
effect of a piece of plane glass.

In order to describe the power of a lens, we say that
it is of such or such a focal length, and, until very
recently, this length was commonly expressed in inches.
For microscopes, telescopes, and optical instruments

generally, the inch scale is still in use; but for spectacles
it has been superseded by a new and more convenient
nomenclature.   Spectacle lenses are chiefly made in
three countries, England, France, and Germany, and
the inches of these three countries are not coincident.
The test cases used by surgeons were until recently
fitted with lenses graduated in French inches; and a
prescription based upon these was often taken to an
optician who supplied lenses graduated in English or
in German inches, and who therefore failed to fulfil the
intentions of the prescriber.   Moreover, as, in the inch
scale, a lens of one inch focal length was the unit of the
series, all weaker lenses, being parts of this unit, were
necessarily expressed in fractions—a system which gave
much unnecessary trouble.   The faults of the inch scale
had long been sources of inconvenience; but the manufacturers of lenses were able for a time to place difficulties
in the way of the adoption of a better system, which
they feared would render their inch-scale grinding tools
obsolete, and would thus entail upon them the expense
of manufacturing new ones.   By the energy of Professor
Donders, of Utrecht, this difficulty was at last overcome;
and we now use, for spectacle purposes, a scale which is
based upon the French metre, and which has a lens of
one metre focal length for its unit.   This unit is called a
dioptric; and, by having a weak lens as the unit instead
of a strong one, nearly all others come to be multiples
of this unit instead of being parts of it, so that they can
be expressed in whole numbers, instead of in fractions.
The lens of one dioptric being No. 1 of the series,

No. 2 is a lens equal to two of the former—that is, it is of double the power of the dioptric, or of half the focal length. It is two dioptrics, and its focal length is half a metre. No. 3 is equal to three dioptrics—that is, its focal length is one-third of a metre. Throughout the series, every whole number expresses the number of dioptrics to which the lens so numbered is equal; and hence, from whole number to whole number, all the intervals are the same. There is a difference of one dioptric between No. 6 and No. 7 ; and there is equally a difference of one dioptric between No. 1 and No. 2. This difference, however, is sometimes too great for practical purposes ; and hence a few quarter and half dioptrics have been added to the lower powers of the scale. These introduce the simple decimal fractions 0·25, 0·50, and 0·75 ; but these fractions are so easily manipulated that they cause no inconvenience. A quarter-dioptric is a lens having a focal length of four metres; a half-dioptric has a focal length of two metres ; and three-quarters of a dioptric has a focal length of one metre and one-third. For ordinary purposes the quarter-dioptrics of the series go up to No. 3 ; and the half-dioptrics to No. 6. For higher powers, such fine divisions are seldom needed, but the uniformity of the scale affords a ready means of supplying them. If we want a lens of six and a half dioptrics, or of eight and a half dioptrics, we have only to add 0·50 as an addition to 6 or 8, and the requirement is at once fulfilled.

In former times, before the principles which should govern the employment of spectacles had been investi-

gated with scientific accuracy, and when people expected
opticians to prescribe spectacles as well as to make and
to sell them, the lenses in customary use were commonly
distinguished by arbitrary names or numbers. A manu-
facturer made, say, twelve different powers, of no definite
relation to one another, and he numbered them from one
to twelve as a matter of convenience. The numbers
really meant nothing, so that the number two of one
maker might be the number three of another, or *vice
versâ*. Sometimes the glasses, especially in the lower
powers of convex lenses, were called by arbitrary names
expressive of their supposed properties, as " preservers,"
" clearers," and such like rubbish. The period of a
definite inch nomenclature commenced about 1860 ; and
the public are now beginning to think of their spectacles
in terms of inches. When a patient says " I have been
using No. 12," he generally means that he has had
glasses of twelve inches' focal length, not some indefinite
number twelve of the pre-scientific period. For those
who have thus become accustomed to think in inches, the
change to the new nomenclature, like all other changes,
may at first be a little confusing ; and there are many
who, on hearing of a lens of five dioptrics, will desire to
refer it to the old scale before they can perfectly realise
its optical value. It will much facilitate the required
translation of the thoughts if we consider that a metre
is equal to 39·337 English inches ; that is, for all practical
purposes, it is equal to forty inches. This means, using
English instead of French inches, that the No. 1 is the
old No. 40, or, more correctly, the old 1-40th. In like

manner, No. 2 of the dioptric scale is the old 1-20th, and No. 4 is the old 1-10th. There are many of these coincidences, the chief of which are shown in Fig. 18, where the horizontal line represents a metre. Besides the three instances already mentioned, it will be seen that three dioptrics coincide, nearly, with 13 inches; 5, with eight inches; 8, with five inches; 10, with four inches; and so on throughout. If we wish to translate dioptrics into French inches, we must consider the metre equal to 36 inches instead of to 40. The coincidences obtained by either method are important, and show that

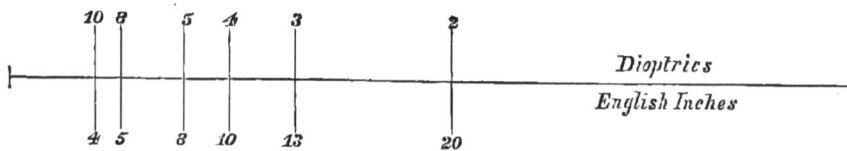

FIG. 18.

the difficulties raised by manufacturers in the way of the adoption of a metrical scale were mostly imaginary; since every glass ground upon the old inch tools has its proper value in the dioptric scale, and all that is necessary is to define and register this value. For the purpose of doing so, Dr. Snellen has invented a very ingenious instrument, called a "Phakometer," which it is not necessary here more precisely to describe, but by means of which it is possible to ascertain the focal length of any lens in dioptrics in a few seconds; and also to ascertain, what is often of great importance, whether the centre of the lens corresponds with its optical axis. We shall see hereafter that inaccuracies in this respect are frequent and fruitful sources of inconvenience.

In writing about lenses, in order to avoid the constant repetition of the words convex and concave, it is usual to distinguish lenses of the former kind by the prefix of the *plus* sign, and those of the latter kind by the prefix of the *minus* sign. Thus, $+ 3 \cdot 0$ signifies a convex of three dioptrics ; and $- 3 \cdot 0$ signifies a concave of three dioptrics.

The materials of which lenses are composed are either glass, or the form of natural rock-crystal which is commonly called pebble. In former times, when the manufacture of glass was less perfect than at present, this material did not long retain complete transparency ; when exposed to atmospheric influences it became more or less turbid from some spontaneous decomposition or re-arrangement of its elements ; and hence pebble, which always retained its transparency, was much more highly prized. For many years, however, a perfectly stable, transparent, and homogeneous glass has been available for the purposes of the optician ; and the only advantages of pebble over glass are such as depend upon the greater hardness of the former, which is therefore not liable to become scratched in use. The refractive power of pebble is also greater than that of glass, so that for equal focal lengths the convexity or concavity of a pebble lens is somewhat less than that of a glass one, and on this account, as well as from the greater strength of the material, the pebble lenses may be made comparatively thin and light. A pebble lens may readily be distinguished from a glass one by its greater coldness to the tongue, pebble being a better conductor of heat than glass; and also by placing it between two plates of tourmaline and holding it up to a window. With the glass, no effect is produced, but with pebble the light is polarised, and rings of colour become visible.

The ordinary advantages of pebble lenses may be more than neutralised if they have not been cut from the original block in the right direction. The material has the curious property of being bi-refringent in one particular direction ; that is, the ray of light passing through it in this direction is split up into two ; and two images of the object from which it proceeds are produced. In order to make a perfect pebble lens, its axis must be at right angles to the axis of double refraction; for otherwise, although the thickness will not be sufficient for two images to be produced, the single image may nevertheless be more or less blurred or bordered. The only security against this for the ordinary purchaser is to buy of an optician of repute, who will be more desirous to supply lenses of the best quality than to make the largest possible number out of a given piece of pebble ; but the matter may easily be tested by the plates of tourmaline already mentioned. If the lens be properly cut, the rings of coloured light will be circular ; if not, they will be elliptical or more or less irregular in shape. A little instrument containing the tourmaline plates is kept by all opticians, and the purchaser may always ask to try his pebble lenses in this way for himself.

CHAPTER IV.

THE FORMATION OF IMAGES IN THE EYE; THE ACUTENESS AND THE FIELD OF VISION; THE BLIND-SPOT.

THE crystalline lens of the eye is an instrument precisely resembling, in its optical effects upon light, a double convex lens made by art ; and, together with the cornea and the aqueous and the vitreous humours, which somewhat assist its operation, it forms what are called collectively the refractive media. For the sake of simplicity, we may regard the collective media as forming a single lens; and, in an ideal or natural human eye, the focal length of this lens is precisely equal to the length of the axis of the eyeball, so that its focus is precisely upon the retina. When this is so, the retina receives a sharply defined image of all objects before the eye which are within the lateral range of its vision, and which are sufficiently remote to transmit approximately parallel rays. Like that of a camera, this image is, of course, inverted ; and it may be seen in the eye of any recently killed animal by the simple expedient of removing a

E

piece of the opaque sclerotic and choroid without inter-
fering with the retina itself. If we hold up the eye so
prepared, we shall see from behind, upon its retina,
an inverted image of the object to which its cornea is
directed. If we use the eye of a white rabbit, in which
the choroid is destitute of pigment and the sclerotic
itself is transparent or translucent, we may see the
inverted image without any preparation.

Recalling what has been already stated, that every
visible object is rendered visible by the light proceeding
from it, whether this be self-luminosity or light derived

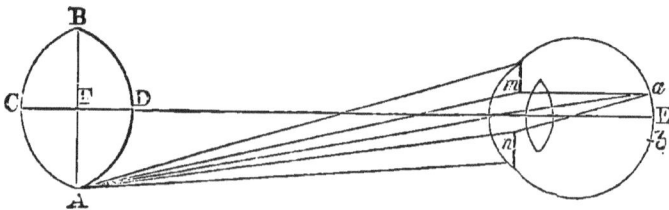

FIG. 19.

from other sources and reflected; and also that every
such body must be regarded as consisting of an infinite
number of luminous points, from each of which light
radiates equally in all directions, so as to form a cone
the base of which rests upon any surface larger than
the point itself, we may easily trace out the formation of the retinal image. In Fig. 19, let us suppose
that the disc A B is a visible object, radiating light from
its surface in the manner described. Of the innumerable luminous points in this surface we will take only
two, the extremities of the diameter A B. From the

point A, a cone of light falls upon the surface of the cornea, but the outer portions of this cone are intercepted by the opaque iris, and, being reflected from its surface, render this surface visible. The central portion of the cone, between *m* and *n*, falls upon the area of the pupil, which in the diagram is somewhat dilated, and passes through it to enter the chamber of the eye. The axial ray of the cone proceeds with but little deviation to the retina, which it strikes at the point *a*, a little above the centre; and the outer rays of the cone are made to converge, by the action of the refracting media, to this same point *a* as their focus. Precisely the same thing happens with the cone of rays proceeding from the point B, which is in like manner brought to a focus on the retina at *b*, a little below the centre. Assuming the disc to be vertical before the eye, with D to the right and C to the left of the spectator, the rays from D would be focussed upon the left side of the retina, and those from C upon the right. The rays from all intermediate points would be brought to foci upon intermediate points of the retina, and the resulting image would be absolutely inverted. That which applies to a disc applies equally, other conditions being the same, to a visible body of any other form; and we only obtain clear retinal images when the light from every point of the object can be brought to a focus upon the corresponding point of the retina in an inverted position.

It was formerly supposed that the blackness of the

pupil, and the invisibility and apparent darkness of the interior of the eye, were due to the absorption of light by the pigment of the choroid. This is not the case ; it being the fact that the eye returns a portion of the light which it receives. The returning rays, however, retrace their paths of entrance, and thus go back to the luminous point from which they issued. Hence an observer, standing in front of an eye and endeavouring to look into its interior, must himself intercept the light by which that interior was previously illuminated; or, if he fails to intercept it, he will no longer be in the path of the rays, and will see nothing. By the simple contrivance of a perforated mirror, fitted with lenses adapted to different conditions, the late Mr. Charles Babbage discovered a means of rendering every detail of the interior of the eye visible ; but his instrument fell into the hands of a practitioner who failed to appreciate its value, and it thus remained unnoticed until, four years later, Professor Helmholtz applied himself to the same problem, and contrived what he called an ophthalmoscope. The instrument of Helmholtz gave only a feeble illumination ; and that of Babbage was shortly afterwards re-invented by Professor Ruete, and has been the occasion of highly important discoveries with regard to the diseases of the eye and their treatment. The ophthalmoscope is now indispensable to the medical practitioner, but its uses are foreign to the subject matter of these pages.

Assuming the eye to be directed straight forward, as along the line E F Fig. 19, it is manifest that it will not

only receive rays of light from the point F, but also
from many points above and below, to the right and
to the left of F, and that these, for a considerable lateral
range, will be brought to foci within the eye. Of
the resulting picture, however, only a small part is
clearly visible ; for the reason that the sense of sight
is far more acute in the central than in the lateral
parts of the retina. If we suppose the disc, A B C D,
in Fig. 19, to be of large dimensions, and the eye to
be directed to its central point F, this would be seen
clearly. The lateral parts of the disc would also be
seen, in the sense that they would be recognised as
existing, and as forming parts of the object of vision,
but they could not be seen with any exactness or
minuteness. If there were anything attractive in any
of these lateral parts, the eye of the observer, in order
to scrutinise this and to see it accurately, would have
to turn away from the central point to the new object
of vision; and the portion of the picture which can
be seen accurately at one time is much smaller than
most people, prior to special observation, would believe.
As far as our sense perception is concerned, the retinal
picture might be exquisitely finished in the centre and
only roughly sketched at the borders, but this is. not
so, and it is the acuteness of sight that fails, not the
clearness of the lateral parts of the image. In this
book, at the ordinary reading distance of fifteen inches,
it is not possible to decipher more than about ten letters
on the same line without a change in the direction of
the eye, although this change is so slight that some

care is needed in order to abstain from making it. In technical language, the whole lateral extent of vision is called the *Field* of vision ; and we are said to see directly with the central part of the retina, and indirectly with the lateral parts. Indirect vision is of great value for many purposes, · and especially for giving us information as to the directions in which it is desirable for direct vision to be exerted. On this account, the indirect is sometimes called the defensive part of the field ; since it gives warning of the approach of large objects, and saves people from being exposed to many dangers. There are certain diseases of the eye in which the outer part of the field of vision is lost, so that the sight is circumscribed as if by looking through a tube ; and, in these cases, although central vision may be good, and the patient able to read small print, there is yet great difficulty in guiding the footsteps and in avoiding obstacles, especially moving obstacles as in the street. There are many persons with contracted visual field who in one sense can see tolerably, and yet who would not be safe in a crowded thoroughfare. The loss of lateral or indirect vision renders them unable to ascertain correctly the relative positions of objects, and entirely conceals from them many which they would require to see in order to guide their steps with safety.

An exceedingly curious example of the effect of contraction of the field of vision was lately related to me by an old gentleman, who had suffered from a malady which produces this effect, but whose remaining central

vision I had been able to preserve by an operation.
With the aid of spectacles he could read such type as
that of this book perfectly, but he was somewhat short-
sighted, and without spectacles even his central vision
was a little doubtful.   Standing one day at the entrance
to the garden in front of his house, he was much puzzled
by the odd movements of two things on the ground ;
things which he thought were two black birds of unknown
species, hopping about and behaving very strangely.
They turned out to be the feet of a market woman
who had brought something for sale, and whose body
was invisible to him so long as her feet were in view.

Much ingenuity has been expended, at various times,
in endeavours to explain how it is that inverted images
upon the retina produce erect vision of the objects
which the images represent.   The most probable of
these explanations is that the nerve elements of the
perceptive layer of the retina, which consist of rods or
cones, having their long axes directed towards the
centre of the eyeball, naturally refer the impressions
which are made upon each of them to the directions
from which the light proceeds.   Thus, the cones of the
upper portion of the retina, being directed downwards,
and receiving the rays of light from below, naturally
refer or trace them to objects at a lower level than the
eye ; while those of the lower portion, being directed
upwards and receiving rays of light from above, as
naturally refer or trace them to objects above the
level of the eye.   In this way, it can readily be understood that the inversion of the retinal image may be

self-correcting, when it is regarded as a source of the sensory impressions which are communicated to the brain.

The central part of the retina, at the posterior pole of the eyeball, is called, on account of its colour, the *yellow spot*, and it is over the yellow spot, that is to say, for the parts of the general image which fall there, that vision is most acute. Within the yellow spot itself there is a central depression in which the acuteness reaches its maximum, and the limit of acuteness appears to depend upon the size of the smallest nerve elements of which the retina is here built up. These elements are cone-shaped, and are of extreme minuteness; and it seems to be essential to the visibility of an object that its retinal image should be large enough to cover at least one cone. The size of the image depends upon two factors, the actual size of the object and its distance; and these determine what is called the visual angle, or the angle formed between two lines drawn from the extremities of the object to meet at the nodal point or optical centre of the eye, a point near the posterior surface of the crystalline lens. In Fig. 20, C being the nodal point of the eye, and A B an object, A C B is the visual angle of that object, and *a b* is the magnitude of its image upon the retina. But the smaller object, A′ B′, which is nearer to the eye, is seen under the same visual angle and forms an image of the same magnitude; while the object D E, which is equal in size to A B, but nearer, is seen under the larger visual angle D C E, and forms the larger retinal image *d e*. In

order that the retinal image may be of the necessary
size to excite perception, the object producing it must,
of course, be seen under a certain visual angle ; and it
has been experimentally determined that square letters,
which have limbs and subdivisions equal in breadth to
one-fifth the height of the letters, and which are at such
a distance that their height is seen under a visual angle
of five minutes, are distinctly legible to the normal eye.
This principle has been applied, by Dr. Snellen of

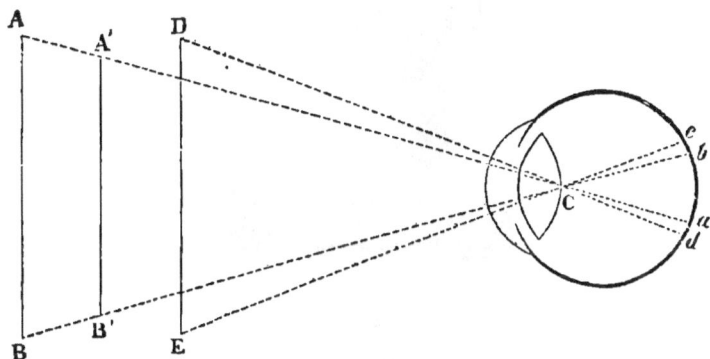

FIG. 20.

Utrecht, to the construction of test-types which afford
a means of testing the acuteness of vision with exactness. His letters are drawn of the proportions mentioned, and of various magnitudes, each distinguished
by a number which indicates the distance, in metres or
parts of a metre, at which the height of the letter will
be seen under a visual angle of five minutes, the breadth
of its limbs under an angle of one minute, and at which
the letter as a whole should be legible. The acuteness
of vision is then expressed by the distance of the

test-types from the eye, divided by the number of the
smallest letter which can be recognised with certainty
at that distance ; the resulting fraction being reduced
to its lowest terms. If the distance be six metres, the
person who can read letters of the corresponding number at that distance has vision equal to 6-6ths, or equal
to 1 ; and this is taken as the normal standard, although
it is not a severe one, and, especially by young people,

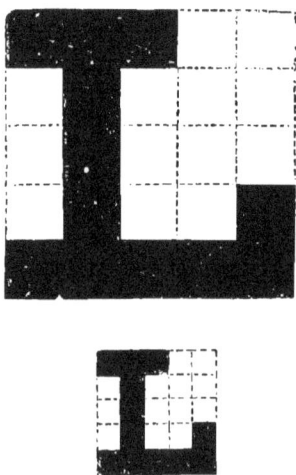

FIG. 21.

is often exceeded. A person who can only read number
9 at the six metres has vision equal to 6-9ths, or 2-3rds of
the normal. One who can only read No. 60 has vision
equal to 6-6oths, or 1-10th of the normal, and so on.
Fig. 21 exhibits two examples of Snellen's letters,
which should be legible at twenty-four and at nine
metres respectively.

The possession of an accurate test of this kind is of

great use to surgeons, who formerly had no means of
estimating vision exactly, or of saying in what degree it
had improved or had undergone deterioration. The impressions of individuals are often utterly untrustworthy,
and nothing is more common than for people who never
had half vision in their lives to represent that they have
always been remarkable for surpassing and unusual
powers in this respect. It is obvious that a man who
cannot tell a cow from a horse across a street may be
perfectly satisfied with himself as long as he has no precise standard by which to measure himself with others ;
and, through the operation of a pardonable human
frailty, while we find numbers of people who admit that
their sight is not as good as it once was, it is rare indeed
to find any one who will admit that his or her sight was
never good at all. In the case of children, as I shall
hereafter have occasion to say at greater length, nothing
is more common than for defective sight to be punished
as obstinacy or stupidity ; and the existence of a visual
defect seems to be about the last thing which ever sug-
gests itself to the blundering self-sufficiency of an
ordinary schoolmaster or teacher. For my own part, I
have long learned to look upon obstinate and stupid
children as mainly artificial productions ; and shall not
readily forget the pleasure with which I heard from the
master of the great elementary school at Edinburgh,
where 1,200 children attend daily, that his fundamental
principle of management was that there were no naughty
boys and no boobies. I think it very desirable that all
parents should test the vision of their children from time

to time ; and I have appended to this volume a page of
print for the purpose, which may be detached and hung
up against a wall.   A child who cannot read every letter
at seven feet distance, has not normal vision, and advice
should be sought for him.   In order to ascertain what
his vision is, he should approach the letters slowly until
he can read them ; and then his distance from them in feet,
divided by seven, will give the fraction that is required.

In addition to the measurement of the acuteness of
central vision, the measurement of the extent of the
field is often required in the investigation of disease, and
is accomplished by means of instruments for the purpose,
called Perimeters, which are made in various forms contrived by myself and others.   The measurement of the
field, however, has no domestic value ; and it is not
necessary to dwell upon it further.

Before leaving the subject of the field of vision, I may
briefly mention the curious gap, lacuna, or " blind spot,"
which it contains, and which corresponds with the termination of the optic nerve within the eye.   This termination,
which is nearly circular in its outline, is absolutely blind,
having even no perception of light ; and perhaps the
most curious thing about it is that, although men had
used their eyes for an unknown number of thousands of
years, the existence of such a blind spot in each of them
was left to be discovered by Mariotte, in the reign or
Charles the Second.   When the existence of the blind-
spot is known, its place is easily determined.   If we take
Fig. 22, and, closing the left eye, look steadily with the
right at the cross to the left of the figure, at the same

time moving the page gently to and fro, we shall soon
find a distance, about eleven or twelve inches, at which
the black circle to the right of the figure entirely dis-
appears from view, the light reflected from it all falling
upon the blind spot. Another method is to close the
left eye, hold out the arms to the front, with the fingers
doubled in and the thumbs in contact with each other

FIG. 22.

and upright. We then look steadily at the left thumb
with the right eye, and slowly move the right arm outwards. When the thumbs are about six inches apart,
the right thumb will disappear from view, to become
visible again as soon as it is either brought back or
moved farther away. Many of the phenomena connected with the blind spot and its vicinity are of great
interest to physiologists, but they have little place in
so elementary a treatise as the present.

It is not only in acuteness of ordinary perception that
the peripheral parts of the retina are inferior to the
central parts, but also, and even especially, in the perception of colour. There are few colours which can be
identified with certainty if the coloured object is held at

one side while the gaze is directed forward, although the object itself may be distinctly seen ; and the difference between black and red in such circumstances soon becomes wholly undistinguishable. A bunch of mixed ribbons affords a good test object for determining this fact experimentally.

CHAPTER V.

I HAVE spoken, in the preceding chapter, of the
formation of images upon the retina by the parallel
rays which proceed from a far-distant object, and under
the supposition that the focal length of the eye as an
optical instrument is identical with the measurement of
its axis, so that the focus of parallel rays falls precisely
upon the retina ; but it is obvious that these conditions
cannot be fulfilled in all cases. There is no necessary
relation between the focal length and the axial length of
the eye ; and our vision is constantly required for near
objects, from which we receive divergent rays.

When the axial and the focal length of the eye are
precisely the same, so that, as shown in Fig. 23, the focus
of parallel rays falls precisely upon the retina, the eye
is said to be emmetropic, from Greek words signifying
that it is " in measure ; " and when this correspondence between the two magnitudes is not exact, the eye
is said to be ametropic, or " out of measure." It is

obvious that the disparity of measure may be in two
opposite directions ; one in which the length of the axis is
less than the focal length, and one in which the length of

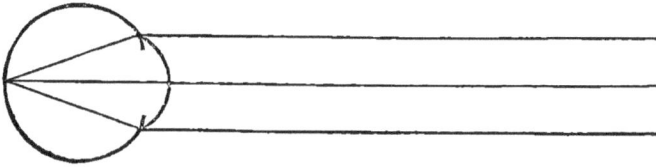

FIG. 23.

the axis is greater than the focal length. The first of these
conditions is shown in Fig. 24, the second in Fig. 25.
The former depends upon flatness of the eyeball from

FIG. 24.

front to back ; and is called flat-eye, or hypermetropia;
the latter depends upon elongation of the eyeball from
front to back, and is the occasion of short sight, or

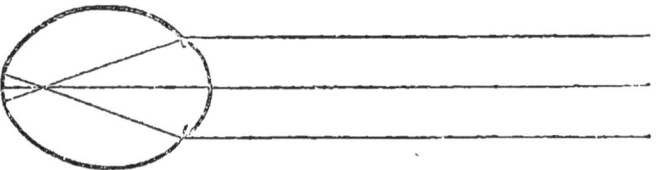

FIG. 25,

myopia. Both of these conditions, or departures from
the right proportion, may exist to a very great degree ;
and, in Fig. 26, the dark line shows the outline of an

emmetropic eye of the natural size; while the dotted
lines show the corresponding dimensions in the most
extreme case of hypermetropia, and in the most extreme case of myopia, in which actual measurements
have been made and recorded.

If the eye were a rigid and passive organ, it is plain
that its influence in refracting the rays of light which
entered by its pupil would be always the same.   In the

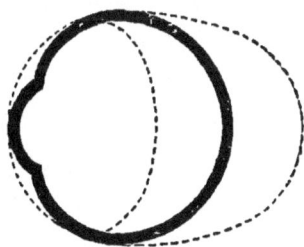

FIG. 26.

emmetropic formation the focus of parallel rays would
be always upon the retina; in the myopic formation,
the focus of parallel rays would be always within the
eyeball but in front of the retina; and in the hypermetropic formation the focus of parallel rays would be
always behind the retina, behind and outside of the eyeball if they could pass through its tunics.   This action
of the passive eye upon light is called its *Refraction;* and
by the refraction of the eye we mean to express whether,
when at absolute rest, it is emmetropic, hypermetropic,
or myopic.   It will be manifest, from Figs. 24 and 25,
that in neither of the latter cases could there be defined
images of distant objects upon the retina.   In the
hypermetropic eye, the rays falling upon the retina

F

would not yet have been united in a focus; and in the myopic eye the rays so falling would have united and would afterwards have overcrossed. In both cases the retina would receive a patch of light, technically called a diffusion circle, instead of the defined image which is necessary to vision.

We have seen already that it is a property of convex lenses to render parallel rays of light convergent, and that it is a property of concave lenses to render them divergent. From this it follows, if we place a convex lens before a hypermetropic eye, that the convergence caused by the lens will enable the eye to bring the rays of light to an earlier focus than by its unaided refraction; and, if we place a concave lens before a myopic eye, the divergence caused by the lens will contribute to bring the rays of light to a more distant focus. For every case of hypermetropia, therefore, there is a convex lens of such a strength that it will exactly suffice, when added to the eye, to place the focus of parallel rays upon the retina; and for every case of myopia there is a concave lens which, when added to the eye, will produce the same result. As shown in Figs. 27 and 28, such lenses at once correct and measure the ametropia: so that we commonly describe the degree of the ametropia in terms of the lens which will correct it. A hypermetropia of two dioptrics is one which requires a convex lens of two dioptrics for its complete correction; and a myopia of two dioptrics is one which requires a concave of two dioptrics for its complete correction.

It is by no means uncommon for the two eyes of the

same person to be of unlike refraction, especially in those
instances in which there is some difference of formation
between the two sides of the face. The difference may
be of every possible kind, one eye being emmetropic
and the other ametropic ; or the two being ametropic in

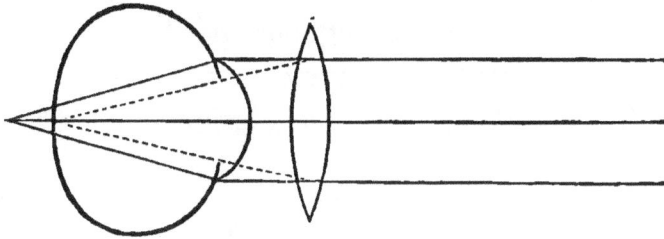

FIG. 27.

different degrees, or even in the contrasted forms ; neither
is it uncommon for the same eye to be ametropic in
different degrees in different meridians. It is almost a
natural formation that the curvature of the vertical meridian of the cornea should be a little sharper than that of

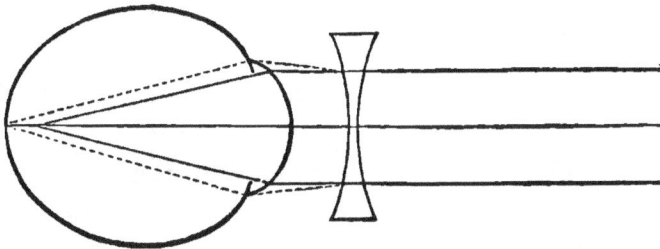

FIG. 28.

the horizontal. When this difference is very slight it is
productive of no inconvenience ; but when it amounts to
a dioptric, or even less, it becomes disturbing to vision,
and constitutes the state known as astigmatism. It is

F 2

obvious that an eye so formed may be myopic or hypermetropic in one direction only; and hence that it may
not see a vertical line as plainly as a horizontal one, or
vice versâ. The test of astigmatism is that all the lines
in such a diagram as Fig. 29, or other similar test object,
are not seen at once with equal distinctness; and, in
high degrees of the defect, one particular line may
appear so confused and blurred as to be scarcely dis-
tinguishable, although the one at right angles to it is

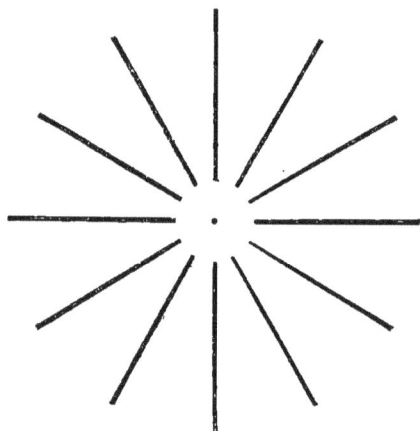

FIG. 29.

clearly visible. It is noteworthy that the meridians of
least and of greatest curvature are always at right angles
to each other ; and the cornea of an astigmatic eye may
be roughly likened to the bowl of a spoon, in which the
curve in the direction of the handle is less sharp than
that transverse to the handle. Astigmatism always
diminishes the acuteness of vision in a marked degree,
and sometimes in a very curious manner. As an example

of this, and to show how easy it may be to mistake a
surmise for a fact, I may mention that a gentleman once
consulted me for what he described as a " periodical
obscuration of vision." I found that he sat in an office
which commanded a view of a large clock-dial on the
other side of a quadrangle. When the hands of the
clock were approximately vertical, he could see them
plainly; but, when they were approximately horizontal,
he could scarcely see them at all. This, which was the
fact, he confounded with his own surmise that he saw
differently at different hours of the day; and hence he
had been induced to read all he could find written upon
the subject of " vital periodicity," and to regard himself
as a curious physiological phenomenon. A pair of cylindrical spectacles at once removed his defect, and restored
his vision to a natural state.   .

Proceeding now to consider the vision of near objects,
and taking the emmetropic eye as a starting point, we
shall see on brief reflection that such an eye, although
capable of uniting parallel rays upon its retina, could
not unite divergent rays if its condition remained unchanged. The divergent rays, such as proceed from all
near objects, require more bending than the parallel rays
in order to unite them in the same distance ; and the
nearer the object, and consequently the more divergent
the rays, the greater amount of bending will be required.
An emmetropic eye, therefore, must possess some power
of increasing its action upon light in order to obtain
clear vision of near objects ; or, in other words, in order
to fulfil the requirements of a function which it is

constantly and almost unconsciously performing. That the eye does possess such a power of variation may be shown not only by reasoning, but in many ways by direct experiment. An emmetropic person, not too old, after looking at the horizon and seeing all its details with perfect clearness, will immediately, and with equal clearness, read a page of small type held at a convenient distance. If he brings the type nearer and nearer to the eye, he will become conscious of a sense of effort; and presently he will reach a point so near that he can no longer overcome the divergence of the rays, and at this point the characters become indistinct. If he now takes a card perforated by a pin-hole, and looks through the pin-hole at the type, thus cutting off the outer or more divergent portion of the cone of rays, and receiving only the axial portion, which is less divergent, he will be able to read at a much smaller distance than before, notwithstanding the greatly diminished illumination which he receives from the page. Again, to borrow an illustration from Professor Donders, if we take a piece of net, and hold it between the eyes and a printed page, we may at pleasure see distinctly the fibres of the net, or the printed letters on the page through the interstices of the net; but we cannot clearly see both at once. When we are looking at the letters, we are only conscious of the net as a sort of intervening film of an uncertain character; and when we are looking at the net, we are only conscious of the page as a greyish background. In order to see first one, and then the other, we are quite aware of a change which occurs in the adjustment

of the eyes; and if the net is very near, and we look at
it for any length of time, the maintenance of the effort
of adjustment becomes fatiguing.

By such observations as the foregoing, and others of
a similar kind, the existence of a power of adjustment
had been established, and the adjustment itself had
received the name of *Accommodation*, long before the
mechanism by which the act was effected was understood.
The actual change produced might be of two kinds,
either an increase in the refractive power of the media,
or an increase of distance between the cornea and the
retina.   It is unnecessary to dwell upon the various
hypotheses which were successively suggested and
abandoned ; and at last, by the labours chiefly of Cramer
and Donders, the act of accommodation was shown to
depend upon an increase in the convexity, and hence
also in the power, of the crystalline lens; and this increase was shown to be brought about by the action of
a muscle which forms part of the ciliary body, and is
named the ciliary muscle, or the muscle of accommodation.   Fig. 30 represents, merely in a diagrammatic
form, the exact nature of the change, in which the
anterior surface of the crystalline lens becomes more
convex, and the pupillary opening smaller.   The left
hand half of the figure shows the parts at rest, the right
hand half shows them as they are when accommodation is
being exerted, and the letters *c m* indicate the position
of the ciliary muscle.

Putting aside the mechanism by which accommodation
is performed, it may be broadly said that the effect of the

change is precisely that of placing an additional convex lens within the eye, and the amount of refracting power which can thus be added is very definite, and admits of easy measurement. For every eye there is a point within which clear vision is no longer possible without optical assistance; and this, which is called the near-point, marks the limit of the power of accommodation. Assuming the eye to be emmetropic, so that it can unite

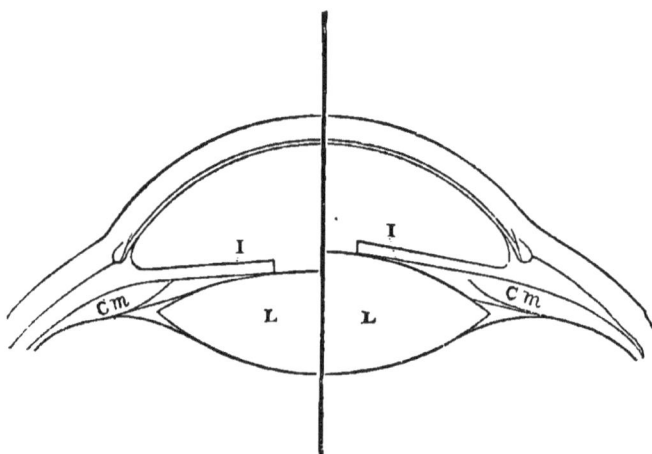

FIG. 30.

parallel rays upon its retina, or, in other words, can see clearly objects which are infinitely distant, let us further suppose that it can also see small objects clearly at twenty centimetres, or one-fifth of a metre, but not at any shorter distance. The effort of accommodation which is exercised in seeing at this near-point produces precisely the same optical results which would be attained by placing within the eye an additional convex lens of the same focal length as the distance from the eye to the near-point. In the case supposed, therefore, the accommodation is equal to a convex lens of five dioptrics. If the near-point were at twenty inches, or half a metre, then the accommodation would be equal to a lens of two dioptrics.

As life advances, the crystalline lens gradually loses its elasticity and becomes more and more rigid ; and, as a necessary consequence of the change, the power of accommodation constantly diminishes, and the near-point recedes farther and farther from the eye. Taking the mean of many observations, we find that at ten years of age the accommodation is equal to a lens of 13 dioptrics, and the near-point is at 3 English inches. At twenty-one, the accommodation has fallen to 9 dioptrics ; and the near-point has receded to 4·5 inches. At forty, the accommodation has fallen to 4·5 dioptrics, and the near-point has receded to 9 inches. At fifty, a great change has taken place. The accommodation is then only 3 dioptrics, and the near-point has receded to 13 inches ; at sixty, the accommodation is only 1·5 dioptrics, and the near-point is at 26 inches ; while at seventy-five the accommodation is wholly lost, the eye is passive, and the near-point is therefore at infinite distance.

The diagram in Fig. 31 shows the effect of declining accommodation in a simple and intelligible form. The strong vertical line, marked with the figure ∞, indicates infinite distance, and the vertical lines to left and right of it indicate dioptrics of refraction. These are numbered consecutively above ; and the figures on the same vertical

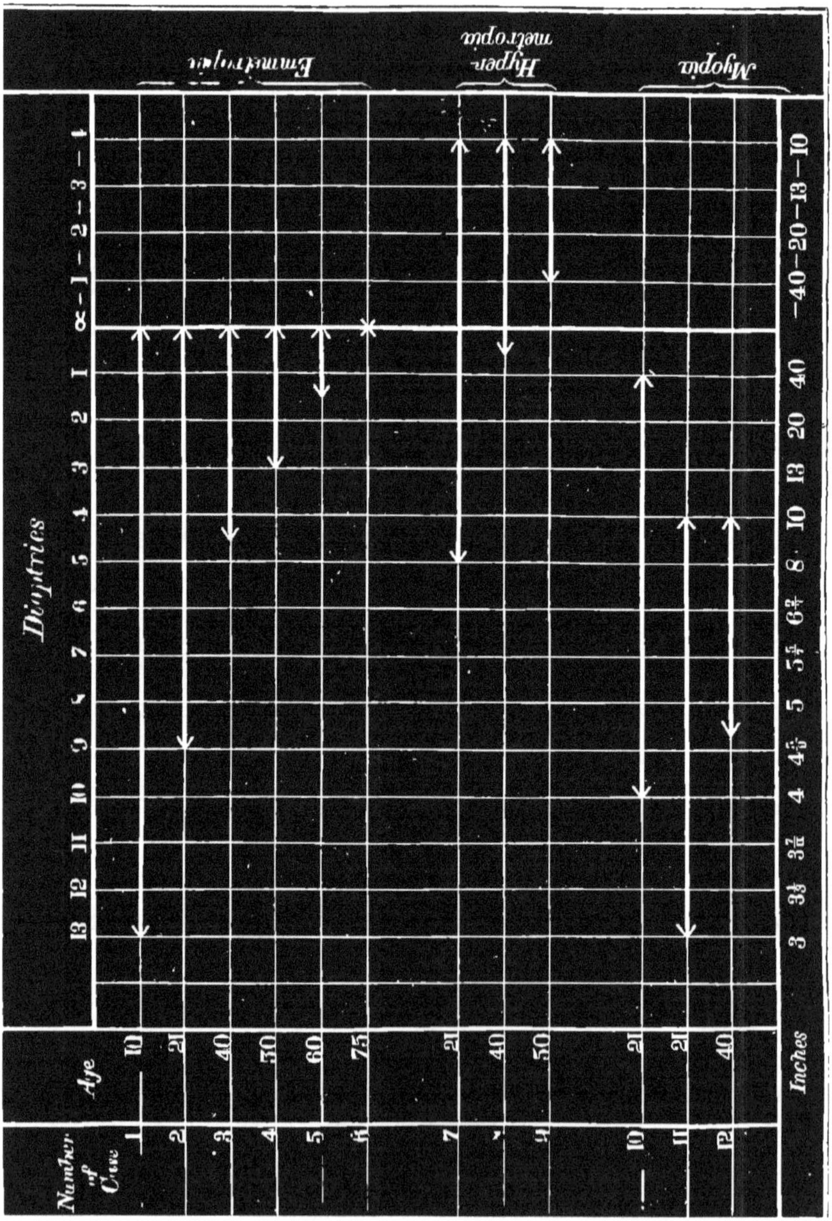
Emmetropia
Hyper-metropia
Myopia
Dioptries
Age
Number of Case
Inches

lines below indicate the distance of the near-point from
the eye, in English inches, for each successive amount
of accommodation. The stronger horizontal lines show
the effort of accommodation in each of several instances;
and the figures to the left of the diagram serve to number
the instances for reference, and to show the ages of the
persons to whom they relate. The first six instances
refer to emmetropic eyes, and exemplify the diminution
of accommodation by age which is described in the preceding paragraph. In No. 1, a person ten years old
brings the near-point to 3 inches by the exercise of
13 dioptrics of accommodation. In No. 2, a person of
twenty-one has 9 dioptrics of accommodation, and
brings the near-point to 4·5 inches. Nos. 3, 4, and 5,
are the other examples; and in No. 6 we have total loss
of accommodation at seventy-five, the far-point and
near-point being the same. The remaining groups
illustrate the accommodation in hypermetropia and in
myopia, and to these we shall return hereafter.

The recession of the near-point, which depends upon
hardening of the crystalline lens, and consequent cur-
tailment of the accommodation, is a phenomenon which
occurs in all eyes, of whatever refraction, and its effects
are known as constituting the condition called *Presbyopia*, or aged sight. Presbyopia is not felt as an inconvenience until the near-point has receded to such a
distance that small objects no longer afford light enough
for comfortable vision; and this generally occurs between
the ages of forty-five and fifty. The change being a
gradual one, it is not possible to say, except arbitrarily,

when presbyopia should commence; but it is held to do
so when the near-point has passed beyond eight inches;
that is to say, when there are only five dioptrics of
accommodation left. It is found, as a general rule, that
not more than half the power of accommodation can be
used continuously, and that any attempt to use more,
except for a few minutes at a time, is attended by
fatigue, which soon compels the abandonment of the
effort. A person, therefore, who has five dioptrics of
accommodation for possible or brief use, will only have
two and a half dioptrics for continued use; or, in other
words, his practical working near-point will not be at
eight inches, but at sixteen. There are many pursuits
which require the object of vision to be nearer to the
eye than sixteen inches; and, unless for large objects or
with abundant illumination, this distance would almost
always be too great to be convenient. We have seen,
however, that the effect of accommodation is precisely
that of adding a convex lens to the passive eye; and
so, when accommodation fails, we can supply its place
by adding the required lens by art. To do this is the
ordinary function of the spectacles which are required
by all people, if their eyes were originally natural, as
time rolls on; the principle on which such spectacles
should be selected is that they should be strong enough
to be effectual; and they should be used as soon as they
are required. Opticians often supply glasses which are
too weak to accomplish what is needed, and which leave
the eyes still struggling with an infirmity from which
they ought to be entirely relieved; while the public

frequently endeavour to postpone what they look upon
as an evil day, and do not obtain the help of glasses
until they have striven hard and fruitlessly to do without
them.   These are important practical errors.   It cannot
be too generally understood that spectacles, instead of
being a nuisance, or an encumbrance, or an evidence of
bad sight, are to the presbyopic a luxury beyond description, clearing outlines which were beginning to be
shadowy, brightening colours which were beginning to
fade, intensifying the light reflected from objects by
permitting them to be brought closer to the eyes, and
instantly restoring near vision to a point from which,
for ten or a dozen years previously, it had been slowly
and imperceptibly, but steadily declining.   This return
to juvenility of sight is one of the most agreeable
experiences of middle age ; and the proper principle,
therefore, is to recognise presbyopia early, and to give
optical help liberally, usually commencing with lenses
of + 1·25, or + 1·50, so as to render the muscles of
accommodation not only able to perform their tasks, but
able to perform them easily.   When, as will happen
after a while, in consequence of the steady decline of
accommodation, yet more power is required, the glasses
may be strengthened by from half a dioptric to a dioptric at a time, and the stronger glasses should at first
be taken into use only by artificial light ; the original
pair, as long as they are found sufficient for this purpose,
being still worn in the daytime.

A popular, but entirely unfounded prejudice, which
exists amongst the public with regard to the hurtful

effects of wearing convex glasses which are too strong,
appears to be traceable to an error founded upon a
curious coincidence. There is a disease of the eye
termed glaucoma, which formerly ended in complete
and irremediable blindness, but which, for twenty years
past, has been cured by operation when recognised sufficiently early. One of the first or even of the pre-
monitory symptoms of glaucoma is a rapid failure of
the accommodation, and hence a frequent demand for
stronger and stronger glasses. At a time when this
disease was very imperfectly understood, opticians saw
many examples of people who came to them for stronger
glasses every two or three months, who were helped by
them for a time, but who soon became totally blind;
and it was not unnatural for them to associate the blindness with the use of the strong glasses. The opinion
thus formed upon a misinterpretation of facts was confirmed by an elaborate article upon spectacles which
appeared, many years ago, in the *Quarterly Review*, and
which did little more than give new life to a variety of
erroneous beliefs. Regarded by the light of modern
knowledge, the facts lend no support to a belief that
strong spectacles can be hurtful; except in so far that
they may produce some fatigue of the eyes and be un-
comfortable, when, in a way presently to be explained,
they disturb the natural harmony between the two faculties of accommodation and of convergence. Remarkable
evidence of the harmlessness of continuous working by
the aid of a single convex glass is furnished by watch-
makers, among whom such work is an unavoidable

condition of their calling, and who appear to me to
enjoy an enviable immunity from eye diseases. It is
exceedingly uncommon to see a working watchmaker
among the patients of the ophthalmic department of a
hospital ; and I entertain little doubt that the habitual
exercise of the eye upon fine work tends to the development and to the preservation of its powers.   It must not
be forgotten, however, that a premature demand for
spectacles may arise from the existence of some diseased
condition ; and hence the circumstance that they were
required at an unusually early age, or that stronger and
stronger powers were asked for, at short intervals, by a
person not originally hypermetropic, showing that the
power of accommodation had declined at more than its
average rate of diminution, would be a reason for dread-
ing the possible approach of glaucoma, and for seeking
skilled advice in time.

The persons who suffer most from popular prejudice
and ignorance on the subject of spectacles are men of
the superior artisan class, who are engaged on work
which requires good eyesight, and who, at the age of
fifty or sooner, find their power of accomplishing such
work is diminishing.   It is a rule in many workshops
that spectacles are altogether prohibited, the masters
ignorantly supposing them to be evidences of bad sight ;
whereas the truth is that they are not evidences of bad
sight at all, but only of the occurrence of a natural and
inevitable change, the effects of which they entirely
obviate, leaving the sight as good for all purposes as
it ever was.   In many shops, in which they are not

prohibited, they are still made an excuse for a diminution of wages; and the result of these practices is
that hundreds of good workmen struggle on perhaps for
years, doing their work imperfectly, when a pair of spectacles would instantly enable them to do it as well as at
any former period. In the present state of knowledge
there is no excuse for rejecting a man's services, or for
diminishing his payment, because he requires spectacles ;
unless it can be shown that, even when he is furnished
with them, his sight is below the natural standard of
acuteness. A person who is emmetropic and presbyopic
does not, of course, require spectacles for distance, but
only for near objects ; and the glasses are placed out of
the way of distant vision by putting them somewhat low
on the nose, so that for distance the eyes look over them.
The same result may be obtained by making the lenses
flat on the top instead of regularly oval.

CHAPTER VI.

SINGLE VISION WITH TWO EYES; CONVERGENCE.

I HAVE hitherto considered the act of vision only as it is performed by a single eye ; but, before proceeding to the effects and the relief of the several forms of ametropia, it is necessary to become acquainted with the mechanism by which the two eyes are so closely combined in the performance of their functions that they act almost like a single organ. The power of the brain to unite into a single perception the two images of an object of vision which are formed upon the two retinæ is called the power of fusion ; and it only exists when the two images are received upon corresponding parts of the two retinæ. This condition, in its turn, is only fulfilled when the eyes are both directed to the same point in space ; and the moment they cease to be so directed, double vision is a result. In Fig. 32, the two eyes, A and B, are directed to the same point C, and the image of C being formed at s, upon the yellow spot of each eye, the two images are fused, and C is recognised as a single point.   In Fig. 33,

on the contrary, the eye A is still directed to C, but the
eye B is directed along the line B B', to the left of C.
In this position, the image of C falls upon the yellow
spot of A as before ; but, as shown by the dotted line,
it.falls upon a part of B which is internal to the yellow
spot, and which therefore, in the natural position of the
two eyes, would receive an image from an object situated

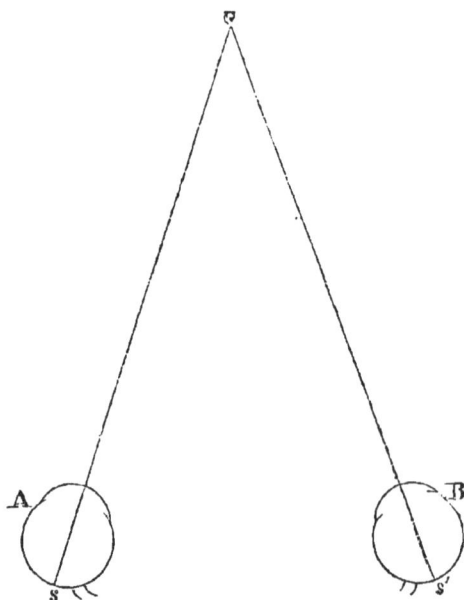

FIG. 32.

at C'.   Hence the eye B refers its image of C to the
position C', or to the right of its real position ; and the
pair of eyes see two points C instead of one only ; one
of them in its natural position, the other outwards or
to the right in the position C'.   The false image, it
will be observed, appears to lie in an opposite direction

to that of the misdirection of the eye which receives it; so that if the eyes squint convergently, or in an inward direction, the double images belong each to the eye on the side on which it appears, and if the eyes squint out-wards, or divergently, the double images are crossed, and each belongs to the eye on the opposite side. The extent of separation of the images is greater, of course,

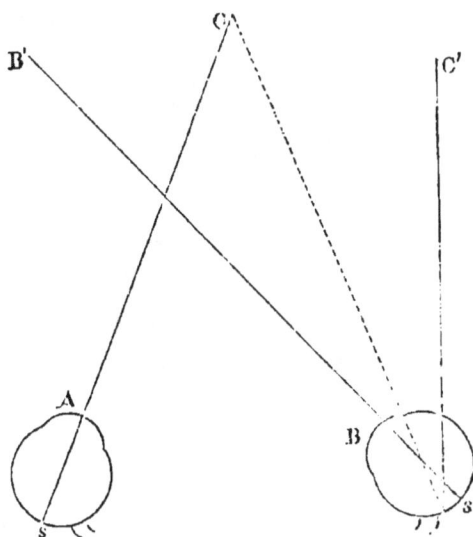

FIG. 33.

as the distance of the object increases; and, for any given object, it depends upon the degree of the dis-placement. When this is very slight, so that the false image falls near to the yellow spot, double vision is very distressing, because the two images are so nearly alike in intensity and clearness that the sufferer cannot readily say which is which, and they render it difficult to guide the steps correctly, or even, sometimes, to touch objects with the hand. When the displacement of one eye is considerable, there is no difficulty in seeing which of them is displaced ; but, when it is slight, this may sometimes be for a time a matter of doubt. The point can always be decided by simple means ; and the question to which eye this or that image belongs is solved by taking a candle flame as the object, and by looking at it with a piece of red glass held before one eye and with the other eye bare. The red image of the flame, whether it be on the right or the left, will necessarily appertain to the eye before which the red glass has been placed.

Besides being displaced inwards or outwards, either eye may be displaced upwards or downwards ; and the maintenance of the correct relative position of the two, in whatever direction they may turn their gaze, is dependent upon the preservation of perfect balance and harmony of action between all the six muscles, the four straight and the two oblique, which were mentioned in the first chapter. It may be said, however, to be more essentially dependent upon the internal straight muscles, shown diagrammatically at I, I, in Fig. 34, than upon any others ; since these are chiefly concerned in the function of convergence, by which the eyes are rolled inwards, and are thus moved, at will, from the position nearly of parallelism which they assume when looking at an infinitely distant object, to that which they assume when directed to some nearer point. If we bring an object of vision slowly near to the eyes of another person, who looks at it as it approaches, we may watch the movement of convergence; and we shall see that, when the effort to look at the near object ceases, the eyes are rolled back to the position of repose, nearly in the middle of each eyelid opening. This rolling back is done by the external straight muscles, E, E, Fig. 34, as soon as the action of the internal muscles is suspended; and any continuance of forced or extreme convergence soon becomes excessively fatiguing. For a short time, however, the eyes may be rolled inwards

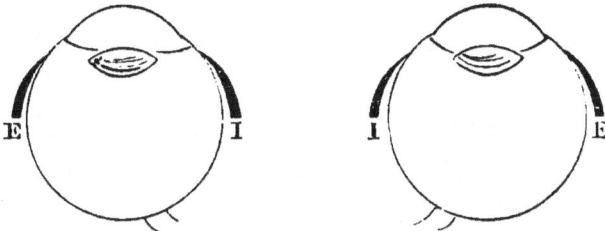

FIG. 34.

in a very considerable degree; so that, especially in young people, they may be fixed upon an object at a distance of only two or three inches from them.

In the emmetropic eyes of a young and healthy person, the two functions of accommodation and convergence go hand in hand, so that they can scarcely be separately performed. Their objects are totally different, but their harmonious co-operation is none the less essential. The function of accommodation has for its purpose the formation of a clear or defined image on the retina of each eye singly; the function

of convergence has for its purpose the fusion of the
two retinal images into a single sensory perception.
But an object at a given distance will always require,
from the same pair of eyes, the same amount of accommodation, and it will always require from them
the same amount of convergence. Say the object is
at ten inches, where it will require four dioptrics of
accommodation. It is evident that it must also require
convergence to such a degree that the lines of direction
of the eyes would intersect at a point ten inches distant
from them ; and further that, if the object were brought
nearer, the accommodation effort and the convergence
effort would increase in an equal degree. Hence it
follows that accommodation for a point, and convergence to that point, become strictly correlated efforts ;
and that one can hardly be accomplished without the
other. The physiology of associated muscular movements is not yet thoroughly understood, and it is by
no means certain whether the connection and inter-
dependence of accommodation and convergence are
natural or acquired ; whether, that is, they are the
results of laws written upon the nervous centres by
which the muscles are called into activity, or only
upon the educational influence of constant consenta-
neous action. However this may be, it is at least
certain that the two functions, in young and healthy
emmetropic eyes, are always performed in unison, and
that they cannot be dissociated without fatigue and
distress. When presbyopia comes on, the tie which
unites them is gradually relaxed ; and, as we shall see

hereafter, in many instances of ametropia it is either
weak or wanting from the beginning.

The power of convergence, with its antagonist, the
power of divergence, or in other words, the strength of
the internal and external muscles, may be conveniently
tested by means of prisms. A prism,
as we have already seen, deflects a
ray of light away from its vertex and
towards its base, as shown in Fig. 7.
In Fig. 35, a ray of light from the
point A, on its way to the left eye
C, in which it would fall naturally
upon the yellow spot S, is deflected by
passing through the prism B. The
ray, being bent towards the base of
the prism, reaches the eye as if it
came from the point A'; and it is
plain that in order to receive the
image of A upon its yellow spot, the
front of the eye must be rolled outwards, in the direction A'. Hence, if
we place before one eye a prism with

FIG. 35.

its base inwards, or towards the nose, we shall produce
double vision unless the eye can be rolled outwards
sufficiently to counteract the refraction of the prism, and
to keep the image still upon the yellow spot; while in
like manner, if the base of the prism is turned outwards,
towards the temple, the eye will be called upon to
execute a convergence movement in order that single
vision may be preserved. In testing the powers of

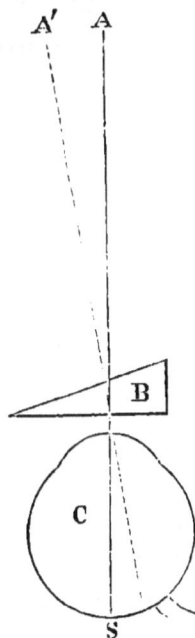

convergence and of divergence, it is better to use a pair of prisms than one only; and the strongest pair which can be overcome, so as to preserve singleness of vision, when their bases are turned inwards, afford the measure of the strength of the divergence function ; while the strongest pair which can be overcome when their bases are turned outwards afford the measure of the strength of the convergence function. The strength of the prisms themselves is expressed by the angle which their sides form with one another.

Although, as we have seen, the original purpose of convergence is simply the accomplishment of fusion, or the maintenance of a single image from vision with two eyes, yet the convergence becomes so closely associated with the accommodation that a mere accommodation effort without reference to fusion will excite and maintain a convergence effort of nearly or quite the same degree. Moreover, in order to exert the accommodation to the fullest extent, it is necessary to exert the convergence also; and this circumstance, as we shall see hereafter, furnishes the explanation of the common form of squint in the great majority of the cases in which it occurs.

The intimate connection between accommodation and convergence may be illustrated by a simple experiment. If emmetropic eyes are furnished with concave glasses, these, by rendering the rays of light divergent, as if they came from a near point, require accommodation to overcome this divergence, even when the gaze is directed to a distant object. The eyes are then accommodating without convergence; and the effort soon becomes wearisome and distressing. If we add to the concave lenses a pair of prisms with their bases outwards, so that the eyes are called upon for convergence as well as for accommodation, the added effort of convergence instantly relieves the fatigue and strain which were due to the maintenance of accommodation alone. In like manner, if we place convex glasses before the eyes of a young emmetropic person, and direct him to look at a near object, upon which he must converge for the sake of single vision, but which calls for no accommodation on account of the presence of the convex glasses, then we find that convergence without accommodation soon becomes as fatiguing as accommodation without convergence. By taking away the convex glasses, so that accommodation is required together with convergence, the fatigue is relieved ; and it may also be relieved by leaving the glasses, but adding to them prisms of proper strength with their bases inwards, so that the demand for convergence is taken away, as well as the demand for accommodation. We learn, from these experiments, that any departure from harmony between the two functions may become a source of fatigue and strain in the use of the eyes ; and the cases are very numerous in which this knowledge is of important practical application in the relief of visual defects.

Assuming the immediate purpose of convergence to be the accomplishment of fusion, the fusion itself greatly increases the usefulness of vision as a source of information concerning the external world. It is by comparison

of the different aspects of any object which are received
by the two eyes, on account of their slightly different
positions with regard to it, that we obtain the idea of
solidity; and, as of late years has been rendered familiar
by the stereoscope, the eyes may even be deceived by
the apparent solidity which is produced by placing
before them two slightly dissimilar pictures. Moreover,
it is by estimating, through the medium of the faculty
called "muscular sense," the amount of effort which
the muscles of convergence and of accommodation are
exerting, that we learn to judge of the distances of the
various objects which surround us. When an object is
very remote, the images of it which are formed upon the
two retinæ can be fused almost without effort ; and, as
the object comes nearer, the necessary effort steadily
increases. If we watch the actions of young children,
especially the extreme uncertainty of their endeavours
to touch an attractive object of vision, we shall speedily
be convinced that the power of estimating distance is in
the human species an acquired and not an instinctive
faculty; although, in many of the lower animals, it seems
to be chiefly of the latter character. If we disturb the
ordinary convergence effort by means of prisms, we can
produce many curious illusions with regard to the size
and the distance of objects, these being, of course, strictly
correlated quantities. Thus, if we look at a near object
through a pair of prisms with their bases inwards, which
render the eyes approximately parallel, as if they were
directed to the horizon, the object will appear to be
larger than it is, because, although its retinal images are

of considerable magnitude, they are fused without convergence effort, as if the object itself were upon the horizon. In like manner, the use of prisms with their bases outwards will make an object appear smaller than it is, because its distance from the eyes, judged by the convergence effort necessary to obtain fusion of the images, will be under-estimated. Several curious experiments of this kind may be made by the simple original or reflecting stereoscope of the late Professor Wheatstone, in which a slight movement of the arms, causing a difference in the angles at which the rays of light from the two figures are reflected into the eyes, will produce great differences in the apparent magnitude or apparent distance of the resulting picture.

CHAPTER VII.

IN the fifth chapter, ametropia has already been defined as a condition in which the length of the axis of the eyeball, and the focal length of its refracting media, are not precisely the same ; and it has been explained that, when the axial length is greater than the focal length, the effect of the disparity is to produce myopia ; while the opposite disparity, or flat eye, is known as hypermetropia. The word myopia is a trivial designation for short or near-sight, derived from the Greek μύω, to shut ; and is based upon the practice, common to the short-sighted, of nipping their eyelids together for the attainment of better distant vision. In a scientific sense the word has not a single recommendation ; but it is understood alike by scientific and by unscientific people, and is probably too deeply rooted in usage to be abandoned for any other.

The ideal or diagrammatic formation of the myopic eye

has been shown in Figs. 25, 26, and 28 ; and the effect
of this formation is to place the far-point of distinct
vision at some measurable distance from the eye.   The
myopic person cannot see to the horizon or to the fixed
stars ; but he can see quite clearly within a certain fixed
limit.   The degree of short-sight is determined solely
by the farthest boundary of vision, and has nothing to
do with the near-point ; which, in short-sighted eyes,
as in all others, gradually recedes as life goes on, on
account of presbyopic changes which diminish the power
of accommodation.

Let it be supposed that a person has not normal vision
at any point more remote than forty inches, but that
within forty inches he can read easily at any point up
to four inches.   The condition of his eyes will then be
precisely what is shown in line ten, in the third group
of Fig. 31.   His vision does not attain to infinite distance, but is limited to a far-point at forty inches ; and
within this distance he has nine dioptrics of accommodation, which bring his near-point to four inches.   If
we supply him with concave lenses of one dioptric, his
distant vision is instantly restored ; because these render
parallel rays of light as much divergent as if they radiated
from a point forty inches distant, and therefore allow
such rays to be brought to a focus upon his retina.   We
say, in these circumstances, that there is myopia equal
to a dioptric.   The concave lenses not only restore distant vision, but they also, when the eyes are directed
to any object nearer than the horizon, consume a
dioptric of accommodation in order to neutralise them ;

and hence, when they are being worn, the near-point is farther from the eye by that dioptric, or is at 4·5 inches instead of at 4.

Myopia of only one dioptric is comparatively uncommon, and much higher degrees are constantly met with in practice. The eleventh line of Fig. 31 shows a much more common degree, in which the far-point or distant limit of vision is at ten inches instead of at forty. In this case, a lens of four dioptrics would be necessary in order to restore distant vision ; and, supposing the person to be twenty-one and to have nine dioptrics of accommodation, the near-point would be at three inches instead of at four, because the whole of the nine dioptrics would be exercised within the ten-inch limit.

Whenever a person who can see perfectly well within a certain limit has less than normal vision beyond this limit, myopia is to be suspected ; and the suspicion becomes certainty if a concave lens restores distant vision nearly or quite to the natural standard. The power in dioptrics of the lens which does this, or which affords the best result, is the measure of the degree of the myopia, which may then be described as equal to so many dioptrics.

Around short-sight several erroneous beliefs have gathered themselves together, and these still retain a vigorous hold upon the credulity of the public. They are, chiefly, that short-sighted eyes are good or strong eyes, that short-sight improves with advancing life, and that short-sighted people need not use spectacles for

reading or other near work, if they can see to accomplish
it without them.

The first of these beliefs, that short-sighted eyes are
good or strong, admits of easy explanation. It rests
upon the fact that short-sighted people can see smaller
objects than others, and that they also see in a fainter
light than others. Both of these conditions depend upon
their power to approximate the object, and upon that
power alone.

The emmetropic person, shown in line two of Fig. 31,
has nine dioptrics of accommodation, which suffice to
bring his near-point from infinite distance to 4·5 inches.
The myopic person, shown in line eleven of the same
figure, has also nine dioptrics of accommodation; but
they bring his near-point from ten inches to three inches.
The former of the two cannot see anything clearly if it
is nearer than four and a half inches, the latter can see
clearly at three inches. In the latter position, from the
larger size of the visual angle and from the approximation
of the object, the myope obtains a retinal image about
one-third larger than, and with more than twice the illumination of, that of the emmetrope. To make the cases
equal, the emmetrope should be furnished with a convex
lens of four dioptrics, and, when he was thus placed
under similar optical conditions, it would generally be
found that he could see the near object better than the
myope. The better sight of the latter is apparent and
not real. If the myopic eye were in other respects
equal to the emmetropic eye in endurance, the power to
approximate an object without the use of a lens might

occasionally be convenient; but, in actual fact, this advantage, if such it may be called, is more than counterbalanced by disadvantages of no uncertain character. The power to see by a comparatively dim light is also dependent upon the power to approximate the object.

The belief that short-sight improves with advancing years has even less foundation. The only respect in which it does so is that the pupil is usually smaller after middle life than before ; and its diminished size diminishes the diffusion circles upon the retina which are formed by the rays coming to it from distant objects. The retinal images, therefore, may appear somewhat more clear than in youth ; but the degree of the myopia, as measured by the lens which corrects it, never diminishes. There is one sense in which it appears to do so, and this is explained by the twelfth line of Fig. 31. Myopic eyes, like all others, become presbyopic, and thus their near-point recedes, and they lose the nearest portion of their range of vision. The person shown in line twelve with a myopia of four dioptrics, and with, at the age of twenty-one, an accommodation of nine dioptrics, would be able, by using half that accommodation, to read comfortably at a distance of about five inches. At the age of forty, as shown in line twelve, when he is deprived of half his accommodation by presbyopia, he will not be able, by the use of half of the remainder, to read comfortably at a nearer point than about six inches. He will have lost a bit of the front of his visual range ; and, in this condition, many people

say, "I am not so short-sighted as I was." Those who
do so are not aware that the measurement of short-
sight is from the far-point, and that this remains at ten
inches, where it always was.

The one advantage which may be conceded to myopia,
when it amounts to as much as three dioptrics, is immunity from the ordinary requirement of spectacles to
read with, in consequence of failing accommodation as
life advances. With a myopia of three dioptrics, which
gives a far-point at thirteen inches, as even a total loss
of accommodation could not place the far-point farther
away than this, so no convex glass can ever be required.
Thirteen inches is a very convenient reading distance,
and no wish to bring a book nearer than this would be
likely to arise. When, however, the myopia is of small
degree, as a dioptric and a half, the loss of accommodation
adds the discomfort of presbyopia to that of the original
defect. A person so situated will need concave glasses
for distance, and convex glasses for reading. I am
acquainted with more than one clergyman who is in
this condition, and who has the two pairs of lenses in
the same frame, the concaves above, and the convex
below. He thus looks up through the former at his
congregation, and down through the latter at his offices
or sermon-book.

When we hear of old people who have "wonderful
sight," so that they can read without spectacles at
seventy or eighty years of age, we may be perfectly
sure that there is nothing wonderful in the matter, but
that such persons are short-sighted to at least three

H

dioptrics, and are enjoying the usual immunity of their state.

If we turn now to the other side of the picture, we find that it is never an entirely favourable, and is sometimes a very gloomy one. The elongation of the eyeball on which myopia depends occurs almost exclusively at the posterior hemisphere, and chiefly in the vicinity of the posterior pole. If we look at the position of the internal straight muscles, as shown at I I, in Fig. 34, we shall see that their strong contraction, so as to produce the degree of convergence necessary in order to give single vision to the short-sighted, must produce also a strain upon the tunics of the eyeball at a point remote from the actual insertions of the muscles, and that this strain will fall upon the posterior pole of the eyeball, and chiefly upon the region between the yellow spot and the entrance of the optic nerve. When approximation of the object of vision, and consequent convergence effort, are habitual and long continued, the tunics of the eye begin to yield to the strain, and the elongation of the globe tends steadily to increase. By this increase, the myopia is rendered "progressive," and progressive myopia is an affection extremely threatening to the sight. The continued strain upon the posterior pole of the eyeball, and the stretching and elongation of the delicate tunics, produce wasting of the tissue of the choroid, sometimes attended or followed by inflammatory changes, and even by detachment of the retina and loss of sight. It is exceedingly desirable that all young persons who are myopic should be carefully watched to see whether their defect is progressively increasing, in order that, if it should present this character, the use of the eyes may be carefully regulated with a view to the progress being arrested. Every myopic eye should be looked upon as a weak organ, capable, indeed, of being preserved in a state of usefulness, but liable to many dangers and mischances from which natural, and even hypermetropic, eyes are comparatively free.

Concerning the causes of myopia, we know little. The affection often descends from parent to child ; and, as it depends essentially upon the shape of the eyeball, it is reasonable to suppose that it may be inherited, and that infants may be more or less myopic at birth. Evidence upon this point would be attainable with difficulty, and only by a careful series of examinations of the eyes of new-born infants with the ophthalmoscope, such as has never yet been attempted upon a sufficient scale. It is possible that the birth condition, or inheritance, may not be one of developed myopia, but only of weakness of the tunics of the eye at the posterior pole, so that they are disposed to yield and undergo elongation as soon as the strain of the internal straight muscles is brought to bear upon them in the act of convergence. When myopia is once established, and unless its influence is speedily counteracted, it provides for its own increase by the effect of convergence effort; and it may attain a considerable degree before its very existence is discovered by parents of only ordinary want of observation. The first attempts to trace myopia to a cause

were made in Germany, where the condition is more
common than in most other countries, by Dr. Cohn, of
Breslau, who set himself, to examine the state of the
eyes of the children in the local schools. He examined
the eyes of 10,060 children, and in this number he found
1,004 who were myopic. He found also, what was far
more important, that the myopia increased steadily,
both in the relative number of cases and in the degree
of the defect, as he ascended the schools from the
elementary to the higher classes. He further found
that there were relatively more cases of myopia, and
higher degrees of the affection, in schools which were
badly lighted ; and he finally referred the facts to the
operation mainly of two causes, defective lighting and
badly constructed desks and forms, both agencies being
alike in causing the children to stoop over their work,
and to bring their eyes as close to it as possible ; both
being alike, that is to say, in compelling the maintenance of a great amount of convergence effort. The
publication of Dr. Cohn's work led to similar researches
elsewhere ; and corresponding facts have been brought
to light in Russia by Dr. Erismann, and in America by
Drs. Agnew and Loring. There is no longer any room
for doubt that badly-lighted and badly-fitted schools
form a great machinery for the development of myopia ;
and it is probable that this machinery, where, as
in Germany, it has for a long time been in unchecked
operation, may have an important influence upon the
form of eyeball which will be inherited by large numbers
of the population. The English school-boards have not

been left unwarned upon the subject. As soon as
Dr. Cohn's work was published, I placed his facts and
figures before the medical profession through more than
one channel; and attention has since been called to
these facts and figures by others. Some two or three
years ago Lord Monteagle brought the subject before
the House of Lords, by a question to the Duke of
Richmond, the Lord President of the Privy-Council;
but the Education Department had very little to say
in reply. The attention of the great majority of school-
boards, in the intervening time, has been chiefly given
to questions which have no doubt seemed to them more
important than the health and eyesight of the children;
but it is not unreasonable to hope that these may, in
their turn, eventually receive some small degree of
consideration. For the prevention of myopia in schools,
there can be no doubt that good and well-placed
windows are essential, and that fittings of judicious
design will be useful; but neither of these will be
effectual, or will prevent children from drooping over
their work, unless the matter receives the constant and
vigilant attention of teachers, and unless the sanitary
state of the buildings, and the time relatively given to
work and to play, are such as to meet the requirements of physical health. It is a curious illustration of
the essentially mechanical character of certain minds
that the progress of the myopia should in Germany
have been referred to the enforced convergence
alone, and that better light and better fittings should
have been put forward as sufficient to bring about

a better state of things. Dr. Agnew, of New York, with more practical knowledge and with deeper wisdom, pointed out that a feeble and easily extensible character of the ocular tunics would be a condition largely depending upon general debility; and that the treatment of this debility by food, tonics, and exercise, as well as by an ample supply of pure and often renewed air in the schoolrooms, a judicious abbreviation of tasks requiring the close application of the eyes, and the use of books printed in bold characters, would be of great assistance in bringing about a much needed reform. The robust faith of the average schoolmaster in the efficacy of what he calls teaching is probably not destined to survive the time when a somewhat better acquaintance with the nature of mental operations will become diffused abroad ; and, in the meanwhile, and with reference to the frequent sacrifice of the physical side of the development of the young, it is not uninteresting to recall the results of an experiment made, some ten or twelve years ago, in the village school at Ruddington in Nottinghamshire, under the direction of the late Mr. C. Paget, sometime M.P. for Nottingham. In this school Mr. Paget introduced a half-time system as an experiment, to which only a portion of the children were subjected, and which amounted to a substitution of garden work for about one half of the ordinary school hours. The children who were so treated were found, after a short period, altogether to outstrip in their school-work those who devoted, or who were supposed to devote, twice as much time to it. The prevention of the increase of short-sight

in schools is less, in my judgment, an affair of desks
and fittings than of careful and judicious sanitation :
for I have no doubt that the optical conditions which
would produce myopia in weakly children would fail to
do so in the robust.   None the less, however, should
these optical conditions, together with the lighting and
the distance of the work, receive a due share of
attention ; although such mechanical matters must
not be expected to supersede the necessity for the
constant supervision of a directing intelligence.

When myopia is fairly established, the principle upon
which it should be treated is very simple.   The great
cause of the increase of the affection, and of all the
dangers to sight which such increase may entail, is the
strain thrown upon the tunics of the eyes by undue and
prolonged convergence ; and therefore this convergence
must be prevented.   The prevention is to be accomplished
only by means of spectacles ; and the object of the
spectacles is not to make the patient see any better, but
to compel him to keep his work farther away.   The
myope will often allege, with perfect truth, that he can
see near things better without the spectacles than with
them, and will often complain of some discomfort from
their use.   It is obvious that his eyes will have been
gradually trained, by necessity, to exert a degree of
convergence in excess of their accommodation ; and that
this, however certain to be ultimately injurious to the
eyes themselves, may have become easy as an accustomed muscular effort.   The spectacles call upon the
muscles of the eyes to work under new conditions, and

these, although more wholesome than those which preceded them, may be irksome as long as they are new ; for this reason amongst others, that the external straight muscles, long accustomed to be almost passive, will be brought into use for the maintenance of fixation at the more distant point of vision. It is obvious that the use of spectacles for reading, writing, and such like occupations, will be most important in early life, before growth is completed, and while the tunics of the eyes, like all other bodily structures, are comparatively lax and yielding. At this time, too, in many instances, the demands of education require a closer application to books than at any subsequent period ; so that, when maturity is attained, the glasses may often be laid aside.

In some instances, the internal straight muscles of a myope are unable to meet the convergence demand required of them for the maintenance of binocular vision, and early give up the endeavour. In these circumstances the eyes are used singly, and objects can be brought close to either of them without producing convergence. Vision with one eye only is always imperfect ; and hence, in the cases referred to, an endeavour to produce fusion of the images by means of glasses which require a smaller degree of convergence should generally be made. When this does not succeed, glasses for near work may be abandoned ; because near vision without convergence has no apparent tendency to produce increase of the myopia. The object of vision is then generally held to the right or left of the median line, in front of the eye which is being employed.

The use of glasses for distant objects will generally at once commend itself to the short-sighted, by the pleasure which they receive from the consequent widening of their visual horizon. Apart from this, glasses for constant wear are greatly to be recommended, especially to the young, as a means of unconscious education. If we consider for a moment what must be the state of a person who has grown up to manhood or womanhood with an uncorrected myopia say of only two dioptrics, that is, with a far-point twenty inches away, we shall perceive the educational importance of the early correction of the defect. An emmetropic person may produce the condition artificially by placing convex spectacles of two dioptrics before the eyes. The artificial myopia thus produced would be deprived of half its inconvenience by the previously acquired knowledge of the exact forms and characters of numerous objects which would be only dimly seen ; but the subject of it would find, for example, that instead of being able to tell the hour by an ordinary drawing-room clock from any part of the room, he would have to approach within three or four feet of the dial in order to perceive dim indications of the hands. He would lose all the play of expression on the faces of persons with whom he was engaged in conversation. I once prescribed glasses to correct the myopia of a lady who had for many years been engaged in teaching, and who had never previously worn them. Her first exclamation of pleasurable surprise, as she put on her spectacles and looked around her, was a curious commentary on the state in which her life had until then been passed. She said, "Why! I shall be able to see the faces of the children!" If we think what this exclamation meant, and if we apply the lesson which it teaches to other pursuits, we shall not fail to perceive that the practical effect of myopia is to shut out the subject of it from a very large amount of the unconscious education which the process of seeing the world involves, and thus to occasion losses which can hardly be made up in any other way. Taken in detail, these losses, the mere not seeing of this or that seeming trifle, may appear insignificant; it is their aggregate which becomes important. A young lady was lately brought to me by her parents, on account of the way in which the effects of her myopia had forced themselves upon their notice during a Continental tour. Two schoolboys were of the party, and they subjected their sister to an unceasing chorus of " Don't you see this?" and "Don't you see that?" and "How stupid you are!" until it became manifest to the elders that a state of things which at home had always been accepted as a matter of course was really a very serious evil. A distinguished man of science, who is myopic in a high degree, and who did not receive glasses until he was nineteen or twenty years old, has often told me how much he had to do in order to place himself upon the same level, with regard to experience of quite common things, with many of his normal-sighted contemporaries; and it will be manifest on reflection that the matters which are lost by the short-sighted, as by the partially deaf, make up a very large proportion of the pleasures of existence. I am accustomed, on this ground, strongly

to urge upon parents the necessity of correcting myopia
in their children ; and I am sure that a visual horizon
limited to ten or even twenty inches, with no distinct
perception of objects at a greater distance, has a marked
tendency to produce habits of introspection and reverie,
and of inattention to outward things, which may lay the
foundation of grave defects of character. Landscape
painters are the only persons to whom a small degree of
myopia can be useful. I once accompanied a landscape
painter on a sketching expedition, and after a time asked
him whether he intended to omit a certain house from
his drawing. He looked up with surprise and said,
" What house ? There is no house there." I at once
understood a curious haziness of aspect with which it was
his custom to clothe distant scenery in his pictures, and
which was greatly admired by many persons who mis-
took it for a skilful rendering of an uncommon atmospheric effect. In fact, it was only what the short-sighted
man saw always before him; and I am sure he must
himself have been greatly puzzled by much of the praise
which he received. Soon after I first published, in the
Practitioner for 1874, a reference to this effect of myopia
upon painting, an endeavour was made to account for
some of the peculiarities of Turner's style by the peculiarities of his vision ; but, as I shall have to explain
more fully when speaking of astigmatism, the views
advanced in this endeavour appear to me to be entirely
erroneous.

In order to display the world it is necessary, generally
speaking, to give glasses which fully correct the myopia ;

that is, glasses of the same number of dioptrics. An eye which is myopic to four dioptrics, and which is furnished with a concave lens of the same power, would, if its accommodation were of natural range, be placed in the position of an emmetropic eye. When at rest, it would obtain perfect images from parallel rays ; and it would require to exert accommodation for divergent rays. We often find, however, that the accommodation of myopic eyes is very defective ; for the simple reason that the muscle by which the adjustment is effected has fallen into a state of weakness, or even of imperfect or arrested development, from disuse. A person who is myopic to four dioptrics can read at ten inches without any accommodation effort at all, and would seldom desire to read at a nearer point ; so that his faculty of accommodation might be suffered to fall into almost complete abeyance. When this has happened, the lens of four dioptrics, which corrects the myopia for distance, cannot be overcome for near vision, and the person cannot read with it at all, or only for a short time and at the cost of much fatigue. Recurring to line eleven, Fig. 31, we have there the diagram of a myopia of four dioptrics in a person twenty-one years old, and with the full accommodation of nine dioptrics. Supposing his myopia to be corrected by a lens of four dioptrics, his state will then be identical with that of the eye in line two of the same figure ; and an exercise of half his accommodation will bring his near-point to nine inches, or nearer than it need be. He will be able to read at the convenient distance of fifteen or sixteen inches by using only two and

a half dioptrics, or less than one third, of his accommodation ; and hence glasses which afforded him complete
correction would be useful and available for all purposes
for which they were required. They would show him
the horizon with perfect clearness, and would enable him
to read easily without undue convergence. But if, instead of the natural accommodation of nine dioptrics, he
had only four dioptrics, the use of half of this would
leave his near-point at twenty inches, a distance somewhat too great ; and, in order to bring his book to fifteen
inches he would have to exert more or less strain, so as
to bring more than half his accommodation into play.
The eyes would then soon become fatigued, and reading
with the glasses would be difficult or impossible. Hence,
when the accommodation of myopic eyes is weak, it is
necessary to give weaker glasses for reading than for
distance. It is not possible to lay down any rule which
will be of universal application, and every case must be
considered and treated upon its own merits; but, in a
general way, it may be said that when a myopia does
not much exceed four dioptrics, spectacles which completely correct it may be used for all purposes, and may
be worn constantly, as if they were parts of the eyes.
When the degree exceeds four dioptrics, it will often be
necessary to use weaker glasses for reading ; and such as
leave two dioptrics of the myopia uncorrected will be
found generally available. Thus, if the myopia be of
six dioptrics, the subject will generally read comfortably
in glasses of four dioptrics, and the same thing will apply
to the higher grades.

Notwithstanding the presence of spectacles, we frequently observe that myopic children have a tendency,
the result of habit, to let their books creep up to the
eyes, or their eyes go down to the work, even although
this entails no small strain of the accommodation. In
fact, the importance of doing away with a habitually
stooping posture is second only to the importance of
doing away with over-convergence ; since the stooping
not only tends to fill the eyes with blood, and to subject
their blood-vessels to a hurtful degree of distension, but
it also contracts the chest, and interferes with the growth
of the body and with the complete aëration of the blood
by breathing.   Parents and teachers should therefore be
careful to see, not only that the glasses are worn, but
also that they fulfil their purpose of keeping books and
work at a distance, which should seldom or never be
less than fourteen inches from the eyes. In order that
this injunction may be fulfilled, it is necessary to prohibit the use of books printed in very small type, and
also to prohibit all attempts to read by defective light,
whether this be natural or artificial. Myopic people,
because they can see at a nearer point than others, can
also see by a smaller degree of illumination ; and hence,
as children, they more frequently than others contract a
habit of reading by twilight, or moonlight, or firelight.
Such a habit cannot be too carefully repressed ; nor can
too much stress be laid upon the principle that the use
of glasses for reading, "not that the patient may see
better, but that he may see farther off," while it is an
absolute necessity in all cases of progressive myopia,

and should always be enjoined during school life, or
during periods of close study, as the only means of
preventing increase of the defect, in consequence of
habitual over-convergence, is not less valuable as an
antidote to contraction of the chest and stooping habits.
It is the more necessary to render this clearly understood, because patients are naturally most disposed to
prize and to use glasses for what cannot be accomplished
without them, that is, for seeing distant objects. They
are often unwilling to use them for near work, alleging,
and for a time with perfect truth, that they can see
better and more comfortably without them. It is not
uncommon, indeed, for short-sighted people, when asked
if they have used glasses for reading, to assume a tone
almost of self-righteousness in their denial of the impu-
tation. They say, " Oh, no, I have never done that!"
and are often greatly exercised in their minds when the
urgent necessity for a total change of their habits in this
respect is explained to them.

In some cases of myopia, we find the defect of shape
of the eyeball complicated and increased by spasm of
accommodation, the effect of which is to make the
myopia appear to be of a higher degree than it actually
is. This happens chiefly in cases in which the natural
union or correlation between accommodation and convergence has survived the production of the myopic
elongation of the eye. We find, in such instances, that
the convergence which is necessary for fusion of the
images of the two retinæ is attended by a corresponding
effort of accommodation ; which, by approximating the

near-point, renders the apparent greater than the actual myopia. The artificial approximation of the near-point calls for a still greater effort of convergence ; and thus a sort of vicious circle is established which tends to the increase of the degree of myopia with excessive and alarming rapidity. In such cases, it is a necessary part of the treatment to restrain the accommodation effort by means of atropia ; but this can only be done under careful medical supervision.

I will pass on now to the opposite form of ametropia, in which the axis of the eyeball is shorter than the focal length of the media, and which has already been described as hypermetropia. A less formidable name, and one which has the advantage of describing the physical condition accurately, is "flat-eyed ; " for hypermetropia appears to depend always upon flatness of the eyeball, and not upon deficient refracting power in the media. The conditions are shown diagrammatically in the second group of lines in Fig. 31, where we have a hypermetropia of four dioptrics at three periods of life : at twenty-one, at forty, and at fifty. At the first period, it will be seen that four dioptrics of accommodation are required in order to bring parallel rays to a focus upon the retina ; the far-point being a mathematical negation on the other side of infinity. Four dioptrics, or nearly half of the total range, being thus used up for parallel rays, it is obvious that the eye, to use a happy expression taken from Professor Donders, begins life with a deficit for all the purposes of near vision. Assuming that it can only use half its power of accommodation continuously, and

its total being nine dioptrics, when the requirements of
distant vision are fulfilled it has only half a dioptric left,
and this would bring its near-point no closer than to
eighty inches, a distance at which small type would be
illegible from the smallness of the visual angle which
it would subtend. The subject, in the case supposed,
would labour under two disadvantages as compared with
a person whose eyes were of emmetropic formation. The
latter would have his accommodation muscle at rest
when the eyes were not directed to some near object,
while the former would never have his accommodation
muscle rested excepting during sleep ; or never, at least,
when the eyes were being used for any seeing purpose.
Secondly, while the emmetropic eyes would see near
objects continuously and without painful effort, the flat
eyes would only see them by an effort of accommodation
which, even if it could be maintained, would be always
irksome and would soon become painful. At the age
of forty, when half the accommodation had been lost
by the operation of time, the remainder would be
barely sufficient for objects at the horizon, and would
leave absolutely no residue available for nearer work ;
while, at the age of fifty, the accommodation would be
insufficient even for the horizon, and the subject would
no longer have clear vision at any distance.

It follows from this condition of things that the presence of flat eyes is generally made known, even in early
life, by the fatigue, pain, and speedy dimness of vision
which attend upon their employment over any matters
requiring close application. The flat-eyed person will

read fairly well for a time, but, after a period varying
with the degree of the defect and with the strength of
the muscular system, there comes to be a consciousness
of effort in the act of seeing. The eyes feel strained,
and the letters become somewhat blurred, and are only
restored to clearness by a distinct and often strenuous
effort. There is an instinctive desire to rest the eyes,
to close them firmly for a moment or two, and, often,
to compress the closed eyes with the hand. After doing
this, a fresh start is made, only to terminate in another
compulsory stoppage after a period shorter than the first.
In some instances, a habit is acquired of unduly approximating the book or other object of vision, insomuch that
the condition has actually been mistaken for the precisely
opposite one of myopia, and concave glasses have been
prescribed for its relief. The explanation of this is that
the nearness of the object increases the magnitude of
the retinal images, and in this way assists the dim sight
in a greater degree than the diminished definition impairs
it. If the subject is compelled to work continuously,
as by the demands of some occupation which cannot be
laid aside, the eyes are apt to become red, blood-shot,
and irritable, and to suffer from obstinate forms of superficial inflammation or irritation. The symptoms increase
progressively with the natural diminution of the power of
accommodation, or they not unfrequently undergo sudden
increase as an effect of the general debility produced by
some form of enfeebling illness. In persons who follow
occupations such as needlework, which throw stress
upon the sight, there is often a perceptible amelioration

on Monday, after the Sunday's rest, and a gradual increase of the symptoms throughout the remainder of the week. The condition thus described was formerly called asthenopia, or weak sight ; and it was well known for many years, and was regarded as incurable, until Donders discovered the nature of hypermetropia, and saw at once how asthenopia might be a mere result of over-strain of the accommodation. Save for the fact that it sometimes depends upon muscular strain of a different kind, it is speedily and permanently curable by the use of convex spectacles of adequate power, which supply the place of the irksome accommodation effort, and render the muscular exertion superfluous. It is a curious fact that many patients, prior to the discovery by Donders, had experienced relief by putting on the spectacles of aged people; but, under the influence of the erroneous notions which formerly prevailed, they were strictly prohibited, under the threat of I know not what dreadful consequences, from availing themselves of the remedy which chance or instinct had disclosed to them. The unfortunates were commonly subjected to a great variety of futile and often of severe treatment; such as bleeding, blistering, and the administration of various reputed remedies; and, in the long run, they were generally advised to become agricultural emigrants, or in some way to seek an occupation in which they would never be required to look critically at near objects.

It is essential to the complete relief of flatness of the eyeball that spectacles should be worn always ; but it will often happen that the patient cannot bear full correction of his defect at first. The instinctive desire for clear images is so strong, that the muscle of accommodation is forced into constant and almost spasmodic contraction to whatever extent may be necessary for seeing the horizon ; that is to say, in the case shown in Fig. 31, to the extent of at least four dioptrics. If we now place a lens of four dioptrics before the eye, the muscle is not at first able to relax itself; and the person has lens and accommodation too. This is equal, of course, to an accommodation of eight dioptrics, which would bring the visual distance to ten inches, and would render all beyond indistinct. In very many cases, the flatness of the eye may in young people be wholly concealed by the accommodation ; in which state it is said to be latent, and is only revealed or rendered manifest by paralysing the accommodation with atropia. Hence, in examining a flat-eyed person for spectacles, it is often necessary to apply atropia in order to determine the precise degree of the affection, or even in order to discover with certainty the fact of its existence ; and, when the effect of the atropia has subsided, it may still be some time before the eye learns that the presence of a convex lens renders the accustomed accommodation effort unnecessary, and acquires the power of laying the effort wholly aside, except when the attention is directed to near objects. It is therefore often necessary to use only a partial correction of the flatness in the first instance, as by means of a lens only one-third or one-half as strong as that which will eventually be required ; and, under this treatment, the eye learns its lesson gradually, and the accommodation is relaxed by degrees. The glass chosen should always be sufficient to relieve, in great part or entirely, the distress which was experienced ; and it will usually be found that some return of the old symptoms will point out the time when the glasses first used may be strengthened, either to the extent of full correction or by another step towards it; the full power sometimes not being used excepting by three gradations. It is manifest, however, that the eyes will be little likely to learn to relax their accommodation unless the glasses are always before them ; since, if they are liable to be every now and then called upon for all the former effort, the causes of the old spasm will be retained in constant operation. In early life, while the accommodation is still powerful, it is seldom or never necessary to give stronger glasses than those which are required to correct the hypermetropia for distance ; but, as years pass on, and presbyopia becomes developed, the failing power of accommodation must be supplied. Thus the subject whose case is shown in Fig. 31, on reaching the age of forty years, as in line eight, even if he had glasses of four dioptrics, to enable him to obtain clear images from parallel rays, would not be able, without further help, to read continuously at any nearer point than about eighteen inches ; and at fifty, (see line nine,) at any nearer point than thirty inches. As soon, therefore, as the presbyopic period of life is reached, we supplement the original correction of the flatness by whatever additional power the presbyopia may require.

A curious effect of hypermetropia is the production of the common internal or convergent squint, an affection which depends upon flatness of the eyeballs in fully 90 per cent. of all the cases in which it occurs. The rationale of squint is that when a flat-eyed child first begins to employ his vision carefully, say at three or four years of age, it is necessary for him to exert a large amount of accommodation. This is more readily done if it is associated with convergence effort ; and hence the latter becomes as habitual as the former. In accordance with a well-known law of the animal organism, muscles which are habitually exercised, without being over-strained, increase in strength and in volume ; and so it befalls that the convergence muscles undergo this physiological change, and, by reason of their increased power and their frequent or constant exertion, they, like the muscles of accommodation, fall into a sort of spasm, from which it follows that the resting or quiescent position of the eyes comes to be one of convergence, say to a point eight or ten inches distant. If this convergence cannot be relaxed, it is obvious that the two eyes will not receive images from any more distant point upon corresponding retinal regions, and hence, for the reasons already stated, double vision of all such objects will be produced. Let it be supposed that the object of vision is at A, Fig. 36, and that the eyes are habitually convergent to the nearer point B, in which position they would have double images of A, which, neither of them being formed on the yellow spot, would neither of them be of the highest degree

of distinctness. If, now, the right eye, R, were to make
a slight rotation outwards, as in the effort of looking to
the right with both, it would receive its image of A upon
its yellow spot, and would see clearly. There is, however, no power to make one eye move alone ; and we can
only look to the right by moving both, an act in which

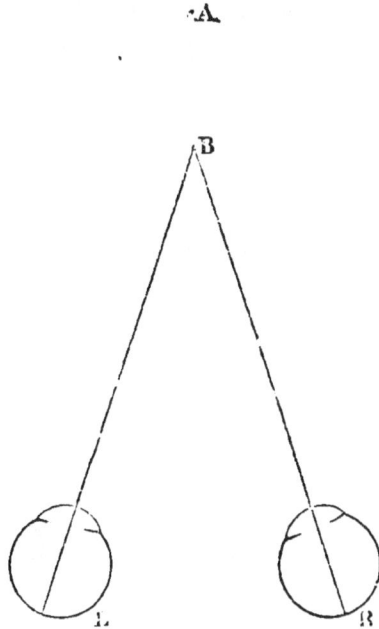

FIG. 36.

the external straight muscle of the right eye, and the
internal straight muscle of the left, work together, and
by a common impulse. Hence, when the right eye rolls
to the right, or outwards, the left eye rolls to the right
or inwards ; and the movements are effected under such
conditions, both eyes starting from a position of convergence, that the same amount of effort which rolls

the right eye a little outwards so as to bring it into the middle of the eyelid opening, rolls the left eye inwards to a much greater degree, so as to make it squint. In consequence of this movement, the image received by the left eye falls upon a comparatively insensitive part of the retina, and is easily neglected by the conscious-ness ; so that the inconvenience of double vision is re-moved by the act of squinting. In Fig. 37, at A, we

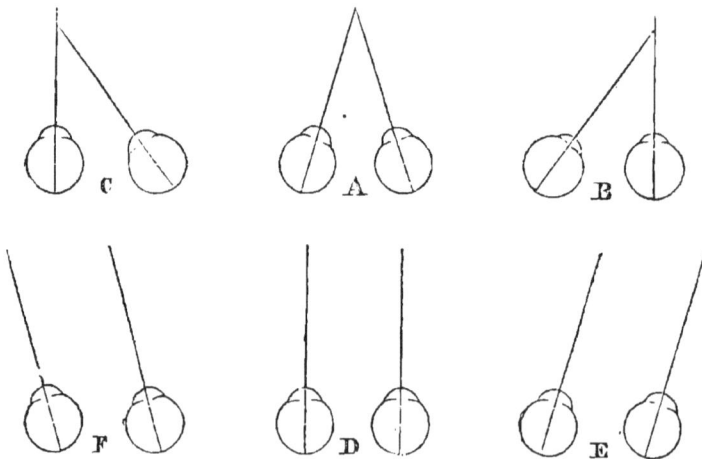

FIG. 37.

have the natural position of squinting eyes, and at B and C we have the positions which they may assume when in use. At B they are looking to the right, making the right the working eye and the left the squinting eye ; and at C these positions are reversed ; but it is evident that in all three figures the relations of the eyes to one another are alike, and the Figures E and F show the effects of the same movements to right and left

when the eyes are natural, as at D, and therefore start
from parallelism instead of from convergence. The right
eye at B has travelled as far as the right eye at E, and
the left eye at C as far as the left eye at F ; and nothing
but the difference in the positions from which they start
produces the difference in the results.

When squinting is first developed it generally affects
the eyes alternately ; that is, the child will sometimes
look to the right, making the right the working eye and
the left the squinting eye, and sometimes to the left,
making the left the working and the right the squinting
eye, in order to see, singly and clearly, an object which
is in front of him. It seems for a time to be a matter
almost of accident which eye may squint ; but, after a
while, either because one eye has better vision than the
other, or because, from some inequality between the
muscles, one is more easily rotated outwards than the
other, we see, as a rule, that one is made always the
working and the other always the squinting eye. As
soon as this occurs, the vision of the squinting eye
usually begins to undergo progressive impairment ; and
the squint should then be cured without delay by an appropriate operation, in order that the impairment may be
arrested and the vision of the squinting eye preserved.
As long as the vision of both eyes is fairly good, a
perfect result may always be obtained by an operation ;
that is to say, the natural position and the proper harmony of movements of the eyes may always be restored.
If, on the contrary, the operation is too long delayed,
until the vision of the squinting eye, and the tone of its

muscles, have alike suffered, nothing better than a coarse result, the removal of a gross and obvious deformity, can in the majority of instances be certainly secured.

Although about 90 per cent. of all the cases of convergent squint are directly produced by hypermetropia, yet the reverse of this proportion does not hold good ; and the hypermetropic persons who squint are probably not more than two-thirds of the whole number, reckoning those only in whom the flatness of the eyeball is distinctly recognisable. The reason why some flat-eyed persons squint, and others do not, appears to depend mainly upon the closeness or laxity of the tie between the functions of accommodation and convergence. When this tie is very close, so that the child cannot accommodate without also converging, squint must almost of necessity be produced ; but, where it is more lax, it is quite conceivable that the necessary accommodation effort may be made, although the eyes are nevertheless kept properly directed by the external muscles in the interests of fusion, that is, for the avoidance of double images.

I have entered at some length into this subject of squinting, because it is only lately that the cause of the affection has been made known, and because the wildest errors are still uttered and believed about it even by intelligent and educated people. Flat eyes are a matter of formation, like the shape or proportions of other features; whence they are often common to several members of a family. Upon this slender basis squint was formerly ascribed to imitation ; just as it has been

ascribed to teething, to fright, and to a countless multi-
tude of other conditions which could have nothing to
do with it. Children have repeatedly been punished for
squinting ; when it would have been as reasonable and
as effectual to punish them for having blue eyes instead
of brown ones. It cannot be too generally known that
children with a certain degree of hypermetropia, and
with a certain closeness of association between accommodation and convergence, must squint by a necessity
of their organisation, as the only means by which they
can overcome the consequences of their ocular defect ;
while those who are differently formed have no cause
or temptation to squint, and would probably be unable
to do so even by any amount of carefully directed effort.
For the sake both of vision and of appearance, every
squint should be subjected to operation ; but the defect
does not afford the smallest ground of objection to the
squinting person as a companion for children, who will
not copy it unless they are compelled to squint by their
own formation. One of the most remarkable instances
of the prevalence of squint in a family with which I
am acquainted came under my notice at St. George's
Hospital, where, in the course of a few months, I
operated upon six sisters and upon their mother.

I have already described astigmatism as a state in
which the curvature of the cornea is different in two
different meridians, which are often vertical and horizontal or nearly so, and always at right angles to each
other. It is necessary to amend this description by
saying, further, that a very small amount of astigmatism

is probably the natural condition of the human eye, and
that the defect must exist in an unusual degree in order
to be disturbing to vision. As contrasted with a portion
of a sphere, the most familiar example of an astigmatic
surface is furnished by the bowl of a spoon, which is
turned upon a shorter radius, in a direction transverse
to the line of the handle, than in a direction continuous
with it. The former is the meridian of greatest curvature ; the latter is the meridian of least curvature. In
the human eye, the meridian of greatest curvature is
most frequently vertical ; the meridian of least curvature
is most frequently horizontal. This rule applies not
only to the slight or customary degrees of astigmatism,
but also to the excessive degrees ; although many instances are upon record in which the ordinary directions
of the meridians have been reversed, and the greatest
curvature has been in the horizontal direction. As a
rule, again, the astigmatic formation is symmetrical or
nearly so in the two eyes, but to this rule we find
numerous exceptions.

When parallel rays of light are refracted by passing
through a medium which presents a convex spherical
surface, all the rays of which the light is composed
become united in a single focal point, except for some
trifling irregularities due to what is called aberration.
Disregarding these, and assuming the refracting medium
to be of circular outline, the light forms a cone between
the medium and the focus, and any section of this cone
in a plane perpendicular to its axis is necessarily a circle.
Hence the diffusion patches which are formed on the

retina of a hypermetropic or of a myopic patient, from rays of light which have not united or which have united and over-crossed, are ordinarily circular ; but those formed in astigmatism are only circular if the retina happens to coincide with one particular point of the refracted bundle. In Fig. 38 A A′ represents a lens which is more strongly curved in the vertical than in the horizontal direction. The result is, that parallel rays $v\,v$, which fall upon the surface of the lens in a vertical plane, are brought to an earlier focus, at v', than the rays $h\,h$, which fall upon the lens in a horizontal

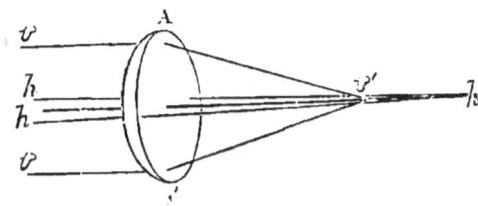

FIG. 38.

plane, and are brought to a focus at h'. If we suppose A A to be the cornea, v' will be the focus of the meridian of greatest curvature; h' the focus of the meridian of least curvature, and the space between the two, $v'\,h'$, is called the focal interval. If we were to intercept the course of the rays by a screen, placed between the cornea A A and the focus v', the diffusion patch would not be a circle, but an ellipse with its major axis horizontal. The rays $v\,v$ would have approached each other more nearly than the rays $h\,h$. At the point v', where the rays $v\,v$ are united, $h\,h$ not being yet united, the diffusion patch would be a horizontal line. A little farther, vv having over-crossed, and hh approaching each other more nearly, the diffusion patch is a smaller ellipse in the same position as before; and this passes into a circle as soon as the over-crossing of vv, and the approach of hh, form equal magnitudes. A continuance of the same process causes the rays to pass into a small upright ellipse, beyond the circle, into a vertical line at the point h', and into a larger ellipse when both the vertical and the horizontal rays have over-crossed

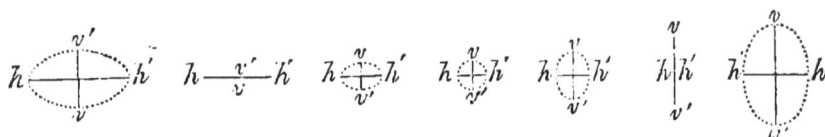

FIG. 39.

each other. These successive forms of the diffusion patches are represented in the series of diagrams which make up Fig. 39, and which are taken from Professor Donders.

When a cornea is so curved as to bring the rays which fall upon it in one meridian to an earlier focus than those which fall upon it in a meridian at right angles to the former, the resulting astigmatism may assume five different forms, which are governed by the position of the retina with regard to the two foci. In the following figures, v always represents the focus of the meridian of greatest curvature, and h the focus of the meridian of least curvature. It will be convenient to assume that the former is vertical, and that the latter is horizontal.

In the first form, Fig. 40, the focus of the vertical meridian is in front of the retina, and that of the horizontal meridian is upon the retina. In other words, the eye is myopic for parallel rays refracted in a vertical plane, and is emmetropic for parallel rays refracted

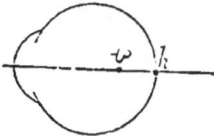

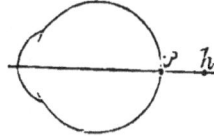

FIG. 40.   FIG. 41.

in a horizontal plane. This is called simple myopic astigmatism.

In the second form, Fig. 41, the focus of the vertical meridian is upon the retina, and that of the horizontal meridian is behind it. In other words, the eye is emmetropic for parallel rays refracted in a vertical plane, and hypermetropic for parallel rays refracted in

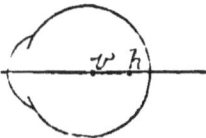
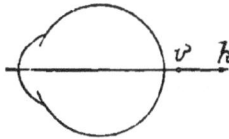
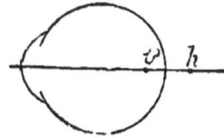

FIG. 42.   FIG. 43.   FIG. 44.

a horizontal plane. This is called simple hypermetropic astigmatism.

In the third form, Fig. 42, the foci of the two meridians are both in front of the retina, but the focus of the vertical meridian is anterior to that of the horizontal. In other words, the eye is myopic for parallel rays refracted in both meridians, but in a greater degree for rays refracted in a vertical plane than for rays refracted in a horizontal plane. This is called compound myopic astigmatism.

In the fourth form, Fig. 43, the foci of the two meridians are both behind the retina, the focus of the vertical meridian being anterior to that of the horizontal. In other words, the eye is hypermetropic for parallel rays refracted in both meridians, but in a greater degree for rays refracted in a horizontal plane than for rays refracted in a vertical plane. This is called compound hypermetropic astigmatism. In the fifth form, Fig. 44, the retina is situated in the focal interval, so that the focus of the vertical meridian is in front of the retina, and the focus of the horizontal meridian is behind it. In other words, the eye is myopic for parallel rays which are refracted in a vertical plane, and is hypermetropic for parallel rays which are refracted in a horizontal plane. This is called mixed astigmatism. Besides these, there are no other forms of regular astigmatism, excepting only that the meridians of least and of greatest curvature may be in any other directions, so long as they are always at right angles to each other. There are certain departures from true curvature which are situated in the crystalline lens, or in structures other than the cornea, and which are called irregular astigmatism, but these hardly admit of being reduced to any general description.

The degree of astigmatism is the measure of the

distance between the foci of the two chief meridians; or, that is, the measure of the difference between the refraction of these meridians, expressed in dioptrics in the usual way. Thus, in a case of simple myopic astigmatism, where the eye is emmetropic for rays refracted in a horizontal plane, with a myopia of two dioptrics for rays refracted in a vertical plane, we say that the astigmatism is equal to two dioptrics. In a case of compound myopic astigmatism, with myopia of two dioptrics for rays in the horizontal plane, and of three dioptrics for rays in the vertical plane, we say that the astigmatism is the difference between the refraction in the two planes, or one dioptric. The degrees of hypermetropic astigmatism are expressed in a precisely similar way; and those of mixed astigmatism by the sum of the two forms of ametropia. Thus, if the eye shown in Fig. 44 had one dioptric of myopia for rays refracted in a vertical plane, and one dioptric of hypermetropia for rays refracted in a horizontal plane, the resulting astigmatism would be equal to two dioptrics.

The correction of astigmatism, that is to say, the equalisation of the refraction in the two chief meridians, is effected by means of plano-cylindrical lenses. We have already seen (*page* 41) that a plano-convex cylindrical lens, with its axis vertical, exerts no influence upon the rays of light which fall upon it in a vertical plane, while it refracts the rays which fall upon it in a horizontal plane. In Fig. 41, which illustrates simple hypermetropic astigmatism, we have an eye which exerts

K

a contrary action, bringing rays which fall upon it in a vertical plane to an earlier focus than rays which fall upon it in a horizontal plane. It is manifest that there must be a plano-convex cylindrical lens of such a power that it will just correct the error of refraction of such an eye ; and this lens, if placed in front of the eye with its axis vertical, will increase the total refraction of the rays in the horizontal meridian until their focus coincides with that of the rays which were previously more strongly refracted in the vertical meridian. In other words, there must be a lens which will bring the focus *h*, Fig. 41, back to the retina, and render it coincident with focus *v*, the position of which will be left unchanged. In like manner, there must be a plano-concave cylindrical lens which will postpone the refraction in the vertical plane of the eye shown in Fig. 40, and will therefore put back the focus *v* to the retina, leaving the position of focus *h* unchanged. A plano-convex cylindrical lens of the proper strength both corrects and measures simple hypermetropic astigmatism, and a plano-concave cylindrical lens of the proper strength both corrects and measures simple myopic astigmatism.

If we look again at Figs. 40 to 44, we shall see plainly that the correction of either of the two forms of simple astigmatism leaves the eye emmetropic as far as its two chief meridians are concerned ; and although such emmetropia is not absolute, on account of the impossibility of exactly correcting the refraction to the intermediate meridians, it is yet sufficient for all

the practical requirements of life. In the compound
forms, however, the correction of the astigmatism still
leaves the eye ametropic; and the correction of the
mixed form may be accomplished in such a manner
as to produce the same effect. Thus, in Fig. 42, the
astigmatism may be corrected in two ways, either by
a plano-convex cylinder which would bring forward
the focus *h* to *v*, and leave the eye highly myopic, or
by a plano-concave cylinder which would put back the
focus *v* to *h*, and would still leave the eye myopic,
although in a less degree. Hence, however the astigmatism is corrected, there must still be a myopic eye;
and the myopia will require a concave lens of the
ordinary spherical kind. In like manner, the astigmatism of Fig. 43 may be corrected by a plano-convex
cylinder, to bring the focus *h* up to *v;* or by a
plano-concave cylinder, to put back the focus *v* to *h*.
Either course would leave the eye hypermetropic, the
former in a much less degree than the latter; and in
either case the hypermetropia would require a convex
spherical lens for its correction. Both the above described conditions are of frequent occurrence, and they
are met by the so-called spherico-cylindrical lenses,
which are ground to be portions of a spherical surface
on one side and to be portions of a cylindrical surface
on the other. Such lenses are made, of course, of any
desired curvature on either side.

Mixed astigmatism, as shown in Fig. 44, may be
corrected either by a plano-concave cylinder, to put
back focus *v* to *h*, combined with a spherical convex

to correct the resulting hypermetropia, or by a plano-
convex cylinder, to bring up focus h to v, combined
with a spherical concave to correct the resulting myopia.
Usually, however, a better way is to employ a bi-
cylindrical lens, that is, one which has a convex
cylindrical surface on one side, and a concave cylindrical surface on the other, the axes of the two being
precisely at right angles. In this arrangement the
concave side should be of the power required to
put back focus v to the retina, and the convex side
should be of the power to bring up focus h to the
retina. The two foci will then coincide, and the eye
will be emmetropic, or nearly so. There is, however,
a difficulty in this mode of correction, arising from the
fact that it will not be accurate unless the axes of the
two cylindrical surfaces are precisely at right angles
to one another. This requirement may be defeated by
the smallest rotation of the lens during the process of
grinding; and its fulfilment can only be secured by
the employment of an optician of adequate skill and
carefulness.

The influence of astigmatism upon the sight is very
considerable, and is exerted in various ways. Its first
and most obvious effect is to produce differences in the
apparent distinctness of equal lines, which are drawn
in different directions; and in this way it produces
indistinctness of some of the linear boundaries of figures,
leaving others clearly defined. Thus, an astigmatic
person, in reading a printed page, may be able to see
clearly the vertical lines which enter into the formation

of many letters, and so, for example, to distinguish an
m from an n. But he would have to place the page
at a different distance, or to alter the accommodation
of his eyes, in order to distinguish the horizontal lines
with equal clearness, and to tell readily an n from an
u. The indistinctness of many boundary lines produces
a corresponding diminution in the acuteness of vision;
and the necessity constantly to alter the adjustment
in looking at the same object produces great fatigue
of the ciliary muscles. Hence, defective sight, coupled
with weariness and aching of the eyes, are the symptoms
of which astigmatic people most commonly complain;
and these symptoms, which may be kept in comparative
abeyance while the accommodation is strong and active,
tend constantly to increase and to become every year
more irksome as the accommodation is curtailed by
advancing life.

The defective vision of astigmatism is primarily for
lines which are at right angles to the meridian of the
preponderating ametropia. In the case shown in Fig.
40, for example, there is myopic curvature in the
vertical meridian ; and in Fig. 41 there is hypermetropic
curvature in the horizontal meridian. The effect of
myopia, that is, of myopic curvature, in the vertical
meridian is to render the eye short-sighted for horizontal
lines ; and the effect of hypermetropic curvature in the
horizontal meridian is to render the eye dim-sighted
for vertical lines. The reason is, that a line is formed,
optically speaking, of a succession of points, from each
of which light is reflected. In the case of a vertical

line, any diffusion circles which may be formed from
the rays which diverge from successive points of the
line in a vertical plane will overlap and conceal one
another in that plane, and will not impair the distinctness of the retinal image ; but any diffusion circles from
the rays which diverge in a horizontal plane will serve
to widen out the retinal image in a horizontal direction,
and will cause the line to appear diffused or spread.
Hence, the eye which has hypermetropic formation in
the horizontal meridian has hypermetropic vision for
vertical lines ; and the eye which has myopic formation
in the vertical meridian has myopic vision for horizontal
lines.   In writing upon the subject, it is necessary to
adopt some uniform system of nomenclature in order
to avoid confusion ; and my own practice is to neglect
the formation and to describe the vision, as the element
which is of the greatest practical concern.   By vertical
myopia, therefore, 1 mean myopia for vertical lines, and
I disregard the fact that the vertical myopia is a result
of faulty curvature in the horizontal direction.   This
method of description has an additional advantage in
the promotion of simplicity ; since the axis of the
correcting cylinder must always be placed in the same
direction as that of the visual defect.

The fact of the existence of astigmatism, although
it may often be inferred from the fatigue and distress
which are experienced in using the eyes, and from the
imperfect relief afforded by common spectacles, can only
be determined by the greater improvement produced by
cylindrical glasses, or by giving the patient an object

of vision which contains equal lines running in different

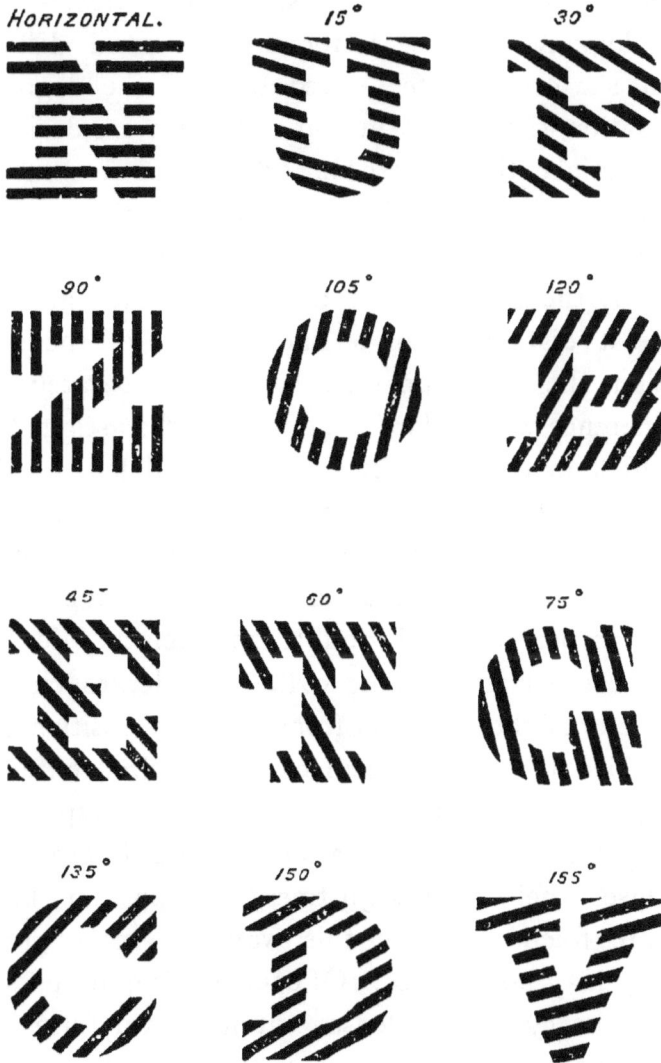

Fig. 45.

directions. Of test-objects of this description I have
found none more convenient than the series of striped

letters shown in Fig. 45, which were designed for the purpose by an American physician, Dr. Orestes Pray. If, on looking at these letters from a good distance, say of six or eight feet, according to the degree of illumination, the stripes in some of them are decidedly more distinct than those in others, the presence of astigmatism is declared ; and the directions of the meridians of least and of greatest curvature will be shown by the directions of the stripes which are most and least distinct. For a line in each of these two directions the refraction must be tested separately, and the difference will declare the degree of the astigmatism, and will show the power of the cylindrical lens required for its correction. The efficacy of this lens, however accurately it may be prescribed, will depend upon the position of its axis ; since a position at right-angles to the proper one will double the defect. For this reason it is usual to make cylindrical lenses circular, so that any slight deviation from the best position may be rectified by turning them to the required extent in the rings of the spectacle frame. When the spectacles are otherwise ready, the patient should himself give the lenses their final adjustment for position before they are fixed ; and in this way the most precise accuracy may be attained. Of course it is quite possible to have the cylindrical lenses cut into oval form after the best position of the axis is determined ; but the London workmen are apt, from want of adequate knowledge of what they are about, to fail in this endeavour, to displace the position of the axis, and to

spoil lens after lens before they succeed.   In the great
cities of America the case is different.   Astigmatism is
not uncommon in the United States, and Americans, as
a rule, will not wear spectacles with circular glasses.
The effect of their determination is that the opticians
of New York and Philadelphia, not to mention other
places, make cylindrical lenses of oval shape with
absolute correctness.

As a cylindrical lens corrects astigmatism, so, of
course, it renders a natural eye astigmatic; and, by
wearing cylindrical lenses as an experiment, it is easy
to feel and realise the conditions of vision which astigmatism produces.   It has been said, and is often supposed, that astigmatism produces distortion of the shapes
of objects, but, in reality, it has no such effect.   If we
place a cylindrical lens in the tube of a magic lantern,
we shall obtain distortion of the images cast upon the
screen ; but this is due to the fact that the rays of light,
after their refraction, travel a sufficient distance to allow
the distortion to occur.   The distance between the cornea
and the retina is not sufficient for this purpose ; and the
only effect of astigmatism is to obscure certain boundary
lines, rendering them hazy and ill-defined.   If we place
cylindrical glasses, with their axes vertical, close to the
eyes, and look through them at any familiar object,
such as a postage-stamp, we shall see no alteration in
its shape; but its lateral boundaries, and the lateral
boundaries of the Queen's head, will look fuzzy and confused.   If we remove the lenses a little way from the
eyes, so as to give the distance necessary for distortion,

distortion will be produced; the object appearing to be
extended in a vertical direction if the lenses are concave, and in a horizontal direction if they are convex.
It is unimportant whether the object looked at is near
or distant. As long as the cylindrical lenses are close
to the eyes, an obscuration of the boundary lines of the
object, in a direction transverse to the axes of the
cylinders, is the only effect observed; but, as soon as
the lenses are held at a distance, alteration of shape
becomes apparent. As the seat of ordinary astigmatism
is in the cornea, what is true of the lens held close must
be true, in a still greater degree, of the astigmatic eye
itself; and hence the suggestion that correctness of
shape in drawing can ever have been disturbed by the
astigmatism of an artist is one which appears to me
to have no sort of foundation in the facts of the case.

CHAPTER VIII.

ASTHENOPIA ; OR WEAK SIGHT.

WE have seen that the vision of the natural eye ranges from infinite distance, or the fixed stars, to a near-point the nearness of which depends upon the amount of the accommodation, which should be about four-and-a-half inches distant at the age of twenty-one, and which becomes progressively more remote as life advances. Within the limits of the range thus specified, or at least within the narrower limits determined by the impossibility of using the whole of the accommodation, the natural eyes can be exerted continuously for indefinite periods ; and it is characteristic of most forms of ametropia that the use of the eyes is not thus free and unrestrained, but is conditioned by limits of time, or by limits of distance, to which natural eyes are strangers. Limitations of distance may be obvious consequences of ametropia ; as when the short-sighted person has his far-point brought within a range determined by the degree of his myopia, and is shut up within a visual horizon of ten or twenty-inches, as the case may be.

Limitations of time are imposed, as a rule, by some
disturbance of the natural harmony between the muscles
by which the eyes are accommodated and directed ; or,
in other words, by want of proper co-ordination between
accommodation and convergence.   Hence, in a large
proportion of cases of ametropia, and in some in which
no ametropia is discoverable, we find, after a period of
use, longer or shorter according to circumstances, either
that the vision becomes indistinct, or that pain is experienced in the eyes themselves, in the regions immediately around them, or even generally over the head.
In many instances the symptoms of distress commence
with indistinctness of vision which, if the effort to see
be continued, leads on to pain.   In many there will
be found sore lids, blood-shot conjunctiva, or some other
sign of superficial irritative or inflammatory disturbance.
In a few the headache is apt to be followed by sickness,
giddiness, palpitation, and other symptoms ; and these
may even be so severe as to occasion a belief that the
patient is suffering from some obscure disorder of the
heart or brain. The foregoing conditions are conveniently
included under the general term of asthenopia, or weak
sight, depending, as they all do, upon some cause which
renders sustained vision impossible.   The word is only
a convenient way of saying that the person cannot use
his eyes for long together ; and it requires to be explained, in every case, by some other which is descriptive
of the precise nature and the apparent cause of the
inability.

    In some instances asthenopia is associated with

manifest defect of vision; in others the vision is little,
if at all, below the natural standard.

It has been stated in a former chapter that asthenopia
was for many years regarded as incurable; and that
Donders, when he discovered the nature and the frequent existence of hypermetropia, was at first disposed
to trace nearly all examples of asthenopia to the strain
which was thrown upon the muscle of accommodation
by its endeavours to correct the flatness of the eyeball.
In an immense number of instances this explanation was
verified by experience, and was found to furnish a clue
to successful treatment by the use of convex glasses;
but the great attention which was soon given to the
matter speedily led to the conclusion that some asthenopic persons were not hypermetropic, but myopic, or
even emmetropic; and it was suggested by Von Graefe
that in these cases the fatigue might be fatigue of the
internal straight muscles by the act of maintaining convergence for some definite distance, instead of being
fatigue of the accommodation. Von Graefe proposed a
division of asthenopia into the accommodative and the
muscular, accordingly as the effort which occasioned
distress was that of accommodation or of convergence.
He supposed that, in cases of the latter class, the internal
straight muscles were the subjects of a peculiar weakness
which he called "insufficiency;" and he suggested tests
by which he thought this insufficiency might be demonstrated, and its degree measured. Later observations
have neither confirmed his hypothesis nor established
the value of his tests; but have tended to show that the

cause of asthenopia, probably in almost every instance, is a want of harmony between the accommodation effort and the convergence effort ; and that the muscular fatigue is due, not to the absolute exertion of either function, but to endeavours to combine the two in proportions which disturb the natural relations between them. When measured by the natural or emmetropic standard, it is evident that in every case of hypermetropia the eyes will be called upon to exert accommodation in excess of their convergence ; and that in every case of myopia they will be called upon to exert convergence in excess of their accommodation. Referring back for examples to Fig. 31, we shall see that the person whose state is shown on line seven would require to exert five dioptrics of accommodation (an amount which would bring the vision of an emmetrope to a point only eight inches away), when his eyes were convergent to a point forty inches away. In like manner, the myope shown on line eleven would require to render his eyes convergent to a point only ten inches distant when his accommodation was absolutely at rest ; and an accommodation effort of one dioptric, which would place the emmetropic visual distance at forty inches, would require from the myope convergence to eight inches. It has been already stated that the correlation between accommodation and convergence is much closer and more complete in some eyes than in others ; as if in some the natural relations had survived the change in the shape of the outward organ, while in others they had yielded to the pressure exercised upon them by the requirements of vision. It seems

that, in cases of the former kind, in which the natural
relation between accommodation and convergence has
been preserved, the combination of the two functions in
unnatural degrees cannot be maintained ; and that this,
rather than the absolute exercise required from either,
is the cause of the asthenopic symptoms. No emmetropic person could maintain convergence to eight or ten
inches, in repose of the accommodation, even for five
minutes ; although nearly every emmetrope of the age
of twenty-one could maintain accommodation *and* convergence for a distance of ten inches even for hours. It
is not surprising that what would be impossible for all
emmetropes should also be impossible for some myopes,
and there can be no doubt that such is the fact. Some
few years ago I saw a very remarkable illustration of
this, which I have already published in detail, but the
chief features of which it is worth while here to repeat.
A young gentleman of good position, who was reading
for honours at his university, suddenly broke down, with
symptoms which were attributed to some form of brain
disease, and was advised to give up his studies and to
go home. After a period of rest, being no better, he
sought advice in London, where the opinion previously
given was confirmed ; and, as a means of affording the
most complete possible rest to his brain, he was advised
to make a voyage to Australia and back. He did so,
and returned in the same condition. He was then considered to be incurable, was told that he must abandon
a career which had been open to him, and a matrimonial
engagement which he had formed. In a word, his whole

life was blighted. Ultimately, he was brought to me, not from any idea that his eyes were at fault, but merely that I might examine their internal circulation with the ophthalmoscope, in order to see whether this examination would throw any light on the state of the circulation in his brain. I found his eyes healthy, but myopic to five dioptrics ; and, on making inquiry into his symptoms, ascertained that they resolved themselves into simple inability to read. As soon as he took up a book, he became giddy ; and the giddiness brought on intense headache, palpitation of the heart, and sometimes sickness. The case was of the simplest kind. The patient had never used spectacles, and, up to a certain point, he had been able to read well and easily ; that is, he had been able to converge to eight inches in repose of his accommodation. When he began to work for honours, and to read eight or ten hours a day, the disproportionate exertion of the two functions could no longer be continued. The convergence muscles gave way ; and then, as the two eyes were no longer directed to the same point, there was double vision. This, in its turn, produced giddiness; and the giddiness produced headache and sickness by disturbing the circulation. The strained muscles which had once given way became prompt to give way again when they were unduly called upon ; and the grave view which was taken of the symptoms by medical men filled the patient with alarm. As soon as he tried to read, the old troubles were brought back by fear and expectant attention. I assured him that he had no brain disease, tried to make him understand his condition, prescribed spectacles to correct his myopia and to diminish his convergence, and told him to wear them constantly and to read in them three times a day for half an hour at a time. He was to report progress in three weeks; and at the end of that time he returned cured. He could read as much as he liked, he was going to be married in the following week, and on returning from his wedding trip was to take up the career which he had fancied closed to him for ever. All these pleasant anticipations were in due time fulfilled, and the cure was permanent and complete.

If, in every case of ametropia, the faculties of accommodation and convergence became disjoined and capable of independent exercise, asthenopia would not occur; and, if these faculties preserved in every case their natural relations, asthenopia would occur in all ametropic persons, but would always be curable by complete correction of the ametropia. Unfortunately, an intermediate condition commonly exists; in which the natural relation is disturbed, but not completely broken through; and it is this which occasions the greatest amount of perplexity. At one time, it was thought that much good might be done in cases of troublesome asthenopia by having convex and concave lenses ground upon prisms; a practice which I introduced to the profession in this country in 1869, by a translation of a work upon the subject by Dr. Scheffler of Brunswick, who, although he was not the originator of the idea, was the first to suggest its systematic application in practice. Starting from the proposition that

L

we can diminish convergence by prisms with their bases
inwards, and can increase it by prisms with their bases
outwards, Dr. Scheffler proposed that prisms should be
used in almost all cases, as means of bringing accommodation and convergence into harmony. The method
was beset by many inconveniences, was of little practical
utility, and was soon abandoned ; so that the use of
prismatic spectacles is now scarcely known except as an
instrument of quackery, and to diminish convergence
effort in a few cases of presbyopia. The true principle
on which asthenopia should be treated is, in the first
place, to prescribe appropriate spectacles for the ametropia alone, and to have them used steadily for two or
three weeks. If, after the lapse of such a time, asthenopia is still troublesome, it is nearly certain that the
patient is exerting either too much accommodation for
his convergence, or too much convergence for his accommodation. In order to redress the balance, it is best
to make the convergence a fixed quantity, say to a point
fifteen inches distant, and to modify the accommodation
to harmonise with it. It is not so much the absolute
accommodation effort that occasions distress, as its disparity with the convergence effort ; and hence we may
always, if necessary, call upon the eyes for a little
more accommodation in order that the disparity may be
removed. To test the nature of the condition, it is necessary to have frames from which the glasses may be shifted
at pleasure, and which will carry two lenses before each
eye. We place in such a frame the lenses which were
first prescribed, and cause the patient to read with them,

at a distance of fifteen inches, until he is tired. As soon as
symptoms of distress appear, we add to the lenses prisms
of about four degrees, with their bases inwards ; thus calling upon the eyes for less convergence. If the previous
accommodation was in excess of the convergence, it
will be left still more in excess, and the distress will be
increased. If the previous convergence was in excess
of the accommodation, the excess will be diminished
and the symptoms relieved. The result, one way or the
other, will be declared in a few minutes ; and it may be
checked by reversing the experiment, and placing the
prisms with their bases outwards, so as to increase the
convergence. In this way, we may find that the previously exerted accommodation was either too much or
too little for convergence to fifteen inches ; and we may
then rectify the disparity by altering the convex or concave lenses. It is manifest that stronger convex lenses
or weaker concaves diminish the demand upon the accommodation ; and that weaker convex lenses or stronger
concaves increase it. By this means it is generally possible to bring the accommodation and the convergence
into harmony by common lenses alone, which may be
obtained everywhere, and to abandon the use of prisms
except for testing purposes. The prisms are not only
heavy and unsightly; but, unless they are cut and fitted
into the spectacle-frames with perfect accuracy, they are
liable to produce disturbing double images in an upward
and downward direction.

In asthenopia of great severity or of long standing,
even when we have corrected every discoverable ocular

L 2

defect, and have brought the accommodation effort and the convergence effort into harmony, we shall sometimes find that the use of the eyes is as painful or difficult as before. For a time this will be so in many cases ; and it is always well for sufferers to understand that their glasses will call upon the eyes to work under new conditions, which, although better than those which they supersede, may yet be irksome as long as they are new. We must not, therefore, reckon too confidently upon immediate relief ; and it is best to insist upon the diligent use of the prescribed glasses for at least a fortnight, before we form a definite opinion about their probable efficacy. After the lapse of that time, if the patient is still complaining, and if we are sure that we have placed the ocular mechanism under the most favourable attainable conditions, we have next to develop the powers of that mechanism by carefully regulated exercise. The patient has often been encouraged, by previous advisers, to " rest the eyes ; " advice which is readily followed on account of the present relief which it affords, but which, several years ago, I described as being always wrong in principle, and always injurious in practice. Rest is necessary in many states of inflammation or disease ; but to rest a muscle which is only weak is the surest way of increasing its weakness, since its nutrition and vigour are always greatly dependent upon proper exercise. We have often to combat a totally groundless fear of blindness, and to deal with organs which have been already " rested " until all their nervo-muscular apparatus has been brought by disuse into a state at once of debility

and of excitability. Such conditions can only be relieved
by careful strengthening of the weakened muscles ; for
which purpose the employment of the eyes must be so
regulated as not to impede nutrition by occasioning
undue fatigue. The best method of fulfilling the required
conditions was first clearly described by Dr. Dyer, of
Pittsburgh, Pa., whose rules for the purpose have proved
so useful, and their value has been so thoroughly established by experience that, in the United States, the
process is commonly called " Dyerising." I will give a
portion of Dr. Dyer's instructions in his own words,
which, on this side of the Atlantic, are less generally
known than they deserve.

" The exercise of the muscles is best accomplished
by reading. The patient is directed to select a book of
good type, but not too absorbing, and to read regularly
with the prescribed glasses three times a day. He must
determine by trial the number of minutes he can read
without discomfort. He may find this to be thirty
seconds, five minutes, ten minutes, or even more. He
must, however, find this initial point. Starting at this
point, he must read regularly, and always with the
glasses. The first reading must not be until one half-
hour after breakfast, the second at noon, the third finished
before sundown. The periods of reading must be regularly increased from day to day. No other use of the
eyes should be allowed. In cases where discomfort
occurs in less than five minutes, the increase should not
be more than one half-minute per day until ten minutes
are reached. In other cases the patient may increase

one minute each day until he can read thirty minutes three times a day without pain. If this can only be done with pain, the patient must be encouraged to persist, notwithstanding the pain ; the surgeon, however, exercising his judgment in not pushing the treatment too rapidly. Should the pain continue from one period to the next, it is evidence that he has gone beyond the maximum of his ability, and that he should fall back to a period at which he can read without discomfort, should regard that as a new point of departure, and proceed as before. As said above, reading is the best exercise; but it frequently happens that the patient is very desirous to write or to sew. This may be attempted when thirty minutes has been reached in the middle period. After the exercise has begun by reading ten minutes, sewing or writing may be tried for ten minutes, and the period finished by reading. From this point I permit an increase of two minutes a day, and a relative increase in the time of writing. This may be gradually introduced into the morning and evening period. I do not consider the treatment completed until an hour and a half has been reached.

"I have found it of great assistance to explain the *rationale* of the treatment to the patient. These cases rarely occurring except in the educated classes, they readily understand their nature and are anxious to assist the surgeon. I tell them that, in reading, pure muscular action is required as much as in lifting a weight; that, through want of use, debility, or some derangement of the system, they have lost the power to exert the

reading muscle without fatigue; that they can strengthen
this muscle and increase its power of endurance by
regular, constant, and systematic exercise, as well as
with any other muscle in the body. The course of
treatment serves to distract the mind of the patient
and restores his confidence in his ability to use his eyes.
He has become discouraged ; he has had the horror of
blindness carefully instilled by friends, and sometimes
by well-meaning physicians, who, not feeling quite sure
of their ground, err on the safe side and prescribe entire
rest. In these cases, the safe side is the wrong side.
When the glasses are procured, and the patient is assured
that there is no absolute disease of the eye as revealed
by the ophthalmoscope, he commences his course of treatment with hope and zeal. The mere fact that he is told
he *must* use his eyes gives him, to a certain extent, the
power to do so."

Dr. Dyer elsewhere lays great stress upon the importance of restraining impatience when improvement is
beginning to be declared. The patient who finds that
he can read for ten minutes without distress is very likely
to go on for twenty minutes, or until pain warns him to
stop ; but to do this is to invite relapse. With an
increase of only one minute a day, the duration of the
treatment would be about three months; and it is better
to submit quietly to this period of modified use, and of
self-restraint, than to lose time at the beginning by
fruitless endeavours to hasten a process which depends
essentially upon the gradual improvement of muscular
nutrition.

CHAPTER IX.

THE phenomena which have been described in the preceding pages have had reference to vision as it deals with the perception of form alone; but it is hardly necessary to say that the information which we receive through our eyes with regard to visible objects is materially enlarged by the different way in which these objects appeal to the sense of colour. The light by which the things around us are illuminated and rendered visible is, in all ordinary circumstances, composed of rays of different degrees of refrangibility and of different wave-lengths; and these differences are perceived, by the great majority of persons, as the causes of many different sensations which we include under the general word colour, and to which a great number of distinctive names have been attached.

It was discovered by Newton that, when a ray of sunlight is made to pass through a transparent prism, the different parts or constituents of the ray undergo unequal degrees of refraction. In order to observe the phenomena in the most perfect manner, the sunlight should be admitted into an otherwise darkened room, through a small circular aperture; and, after refraction, should be received upon a white surface. Instead of the ray as a whole being merely displaced towards the base of the prism, in the way already described, so as to produce a circular spot of white light in a position different from that in which this spot would appear if the prism were absent, the light is spread out into a horizontal bar, called the prismatic spectrum, which is . again divided vertically into bands of colour, successively red, orange, yellow, green, blue, indigo, and violet, and which contains also invisible rays, called " ultra-" red, and " ultra-" violet, beyond its apparent boundaries. Of the several bands, the red portion is the least, and the violet portion the most, diverted from the original track of the ray ; so that the former is said to be composed of the least refrangible rays, and the latter of the most refrangible rays. Investigations of much later date have appeared to show, first, that differences in refrangibility are due to, or at least are associated with, differences in the lengths of the undulations or waves by which the phenomena of light are believed to be produced, the longer waves being less refrangible, or less liable to be diverted from their original course, than the shorter ones ; and, secondly, that four of the seven apparent colours of the spectrum are compound, the results of the admixture, or of the overlapping of the margins, of the three simple colours. With regard to these simple colours, there has not been entire agreement among physicists; but the latest researches point to red, green, and violet, as those from which all others are produced. It is necessary to remember, in dealing with this subject, that it has reference to light alone; and that no inferences can be drawn from observations made upon pigments, which are always compound colours, even when they produce to the eye the effect of simple ones. The colours of all the objects in nature are due to the different degrees in which they absorb or reflect the several elements which enter into the formation of white or solar light; and it will be manifest that an infinite variety of compounds may be produced by the breaking up and disturbance of light waves by the crossing or admixture of others of different length, or of the same length but not vibrating in unison. In the normal eye, the retina is capable of being excited to sensation by light waves of a great variety of lengths and vibrations, and is therefore susceptible of a great number of different shades and intensities of colour, for the due differentiation of which a somewhat formidable vocabulary has been invented and is constantly being enlarged. But there are some persons whose eyes are blind to colour; that is to say, whose retinæ are unable to distinguish between different wave-lengths, but receive the same sensory impression from them all. These persons see no colour; and only perceive through their eyes the forms of objects, and differences in the absolute amount of light reflected from their surfaces, without reference to its quality; differences, that is, of light and shade, but nothing more. Such truly colour-blind

people are very rare; but there are others, far more
numerous, who are blind to one colour only, sometimes
entirely, sometimes only partially. Of these, some are
blind to red, some to green, and a few to violet. For
the sake of illustration I will assume what is not strictly
the case; namely, that the three simple colours enter
into the formation of white light in equal proportions.
It is manifest, on this assumption, first, that the person
who is blind to one colour will perceive all white objects
under one-third less illumination than they present to
normal eyes; secondly, that objects will appear to him
less illuminated in proportion as they reflect a larger
amount of the colour to which he is blind and less of
the others; and thirdly, since there is no object in nature
which absorbs the whole of any of the primary colours,
and reflects only a dual combination, that every colour
will assume an unnatural aspect in the eyes of the
persons who are blind to one only, and who therefore
lose one of the elements which enter into and modify all
compound colours in the eyes of the normal sighted.
The question of colour-blindness has been brought into
great prominence by the demands of railway and steam-
boat signalling; since it is necessary to use red and
green lights or signals for these purposes, and the difference between red and green is that with regard to
which the partially colour-blind most frequently fall
into error. Blindness to red and blindness to green are
the most common forms of the defect; and either of
these would involve a liability to mistakes by which
many lives might be sacrificed.

To determine the fact of colour-blindness is less easy
than might be supposed ; and many inquirers have
adopted methods of so faulty a kind that the conclusions
based upon them are entirely valueless. For example,
ignorant people, or children, have been asked to name
different colours at sight, and have been set down as
colour-blind when they named them incorrectly ; although the errors, in many cases, were afterwards shown
to have been due only to an imperfect acquaintance
with the chromatic vocabulary. A test which at first
seems more trustworthy, although it is not so in reality,
is to place before an intending signal-man, or outlook-
man, red and green discs, or lights, alternately, as he
would see them when on duty, and to ask him, again
and again, which is which. The source of error here is
that the coloured objects are presented to the sight
under the same conditions of illumination, so that, to
the colour-blind, one will look brighter than the other.
The colour-blind person will see *a* difference between
them, although not *the* difference ; and thus, if he
happens to be right the first time, he will keep right all
through as long as the same conditions of illumination
are preserved. Suppose the man thus tested to be blind
for red. He will not see any difference of colour
between the red disc and the green one, but the former
will look darker to him than the latter, because he will
be blind to the greater portion of the light which is
reflected from it. In the same way, a red lamp will
shine with a much fainter light than a green one, supposing the flames behind the two glasses to be of equal

intensity. The perception of this difference in the
degree of illumination, as a matter of comparison
between two signals, would not in the least assist him
to discover whether a single signal, seen under different
conditions of illumination, was of one colour or the
other; and the test is therefore worse than valueless,
because it is misleading.

The only way in which the fact and the nature of
colour-blindness can be certainly determined is one which
has been devised by Professor Holmgren of Upsala ; and
which is as much superior to other methods in simplicity
as in accuracy. The examiner is provided with a large
number, about 150, skeins of Berlin wool; the collection
including all the chief colours and several shades of
each. The skeins are placed in a heap on a table
covered with a white cloth, and in good daylight. The
examiner selects two or three skeins of decided colours
from the heap, and places them aside ; and the person
examined is then told to select and place beside them
other examples of the same colours. There is no question of nomenclature, but only of selecting matches, not
absolute matches, but other skeins which convey to the
eye under examination a similar impression to those
selected as tests. If the person examined matches correctly, his colour-vision is assured ; if he makes mistakes, those mistakes demonstrate not only the fact of
his colour-blindness but also its nature ; whether, that is,
he is blind to colour absolutely, to red, to green, or to
violet.

The limits of this treatise will not allow me to pursue

this very interesting subject to its proper limits; but readers who desire further information with regard to it may obtain it from a volume on the subject by Dr. Joy Jeffries, of Boston, which is published in London by Messrs. Trübner. Dr. Jeffries has for many years been engaged in original investigations upon the subject, and has also largely availed himself of the works of Professor Holmgren and other authorities. He has brought together a number of facts showing the frequent hereditability of colour-blindness, and also the curious preponderance of the male sex among the subjects. Taking the mean of trustworthy observations in all countries, it would appear that rather more than four per cent. of males, and not more than a fifth per cent. of females, are the subjects of some form or degree of the defect.

Colour-blindness properly so-called is a matter of formation, and cannot in any way be modified by training or practice. It would be no more possible to enable a colour-blind person to recognise colour, by any amount or kind of teaching, than it would be possible to enable a totally deaf person to recognise sound. The organ of the faculty is wanting or inactive, and the faculty itself is absent. Many well-meaning but in this matter ignorant people have talked and written about educating the sense of colour, which can no doubt be done when it is present, but not when it is absent; and their efforts in several instances have been manifestly directed to the not very difficult task of teaching the names of colours to children who were always perfectly well able to distinguish them, but who blundered when asked what

they ought to be called. A remarkable intellectual
phenomenon has lately been exhibited in a letter ad-
dressed to the *Times* paper by a gentleman whose name
escapes my recollection, but who described himself as
an inspector of schools, and who set forth that he never
had been able to establish the fact of the existence of
colour-blindness in children under seven years of age.
He apparently wished this inability on his part to be accepted as evidence that all children are born with correct
colour perception, and that colour-blindness is a trick,
or habit, or incapacity, acquired after the mysterious
age of seven years has been passed. It requires unusual
intelligence in the subject, and great patience and tact
on the part of the examiner, even to determine the
degree of acuteness of ordinary vision by letter tests in
a child under seven ; and to ask little children questions
about colours, or to attempt to test their colour vision
except by Holmgren's method, is merely to add one
more to the many possible methods of wasting time
which the ingenuity of former generations has discovered.   Even Holmgren's test, which requires practice
and aptitude in its application to uneducated adults,
would require these in far greater proportion, if any
attempt were made to introduce it successfully within
the walls of an infant school.

Besides true or congenital colour-blindness, there are
forms of incapacity to distinguish colour which may be
produced by disease.   In jaundice the transparent
media of the eye may become tinged with yellow, and
all objects of vision will then appear as if seen through

yellow glass. The drug santonine produces yellow vision in certain circumstances. In some forms of disease of the brain or spinal cord, the loss of colour-vision is an early and a very important symptom, which is often perceived because it is often looked for ; and it is highly probable that the same defect occurs, more frequently than is supposed, in connection with transient forms of brain disturbance, and passes away again as these are relieved or disappear. In certain diseases of the retina colour-blindness is observed as a symptom ; and it is manifest that the natural colour perception is capable of being disturbed, either permanently or temporarily, by a great variety of conditions. The letter from the inspector of schools, to which I have just referred, was followed or preceded by another from a hospital physician, relating the case of a costermonger who had failed, to his own great amusement, to distinguish colours when first tested, but who, by perseverance, and by the diligent care of the sister of the ward, was taught to recognise them before he was discharged. The inferences suggested were that colour-blindness, at least in this case, was merely another word for ignorance depending upon the imperfect cultivation of a faculty, that the man could not distinguish colours because he had not been taught to distinguish them, and that those who can distinguish them are indebted, more or less, to education for their capability. The narrator apparently failed to perceive the bearing of what was incidentally mentioned as part of the history of the case; namely, that the patient was admitted into hospital on account

of a "nervous affection." There can be little doubt
that he was rendered temporarily colour-blind by some
disturbance of the circulation in his brain, that he
recovered his colour-perception as he recovered his
health, and that his recovery was promoted by judicious
cultivation and exercise of the impaired faculty. The
very fact of the amusement which was afforded to him
by his own mistakes is sufficient to show that he had
been able to perceive the differences between colours
at some former period ; for, if not, he would not have
been sufficiently aware of the marked nature of these
differences to have been amused at his mistakes when
he was told of them. If a man has once seen red, it is
intelligible that he should be amused when he is told
that he has mistaken it for some other colour, or some
other colour for it ; but, if he has never seen red, and
has no idea what it is like, such a mistake will appear
to him to be perfectly natural, and will afford no matter
for amusement or even for surprise. The difficulty
would be to avoid the mistake, not to fall into it ; and
this part of the case in question would alone be sufficient
to afford conclusive evidence of its real character.

If we reflect upon the nature of visual impressions,
and even upon the little certainty afforded by verbal
descriptions that different people see things precisely in
the same manner, we shall readily recognise that colour-
blindness, which is a mere negation, the mere non-
recognition of a particular quality in a particular
manner, may easily escape detection unless it is carefully
looked for. What the world in general calls red or

M

green or violet presents, to the sense perceptions of
the colour-blind, an appearance of some sort, and
this appearance he comes to associate with the sound.
Many of the colour-blind have remained for long
periods unconscious of their defect, until its existence
has been established by accident ; and some have shown
great ingenuity in the recognition of colour by other
qualities of the coloured surfaces. It is very important
that a defect, which disqualifies for certain positions in life,
should be recognised early ; and it is therefore desirable
that parents should test the colour-vision of their
children from time to time, not necessarily in any
systematic or obvious way, as a set examination to
be gone through, but by observing whether they are
capable of matching colours correctly. The principle
of Holmgren's method, which is the only sound one,
may be applied sufficiently for all domestic requirements
without having recourse to any other coloured objects
than those which every ordinary household will supply ;
and any mistakes which seem to indicate defective
colour-vision should not at once be made subjects of
comment, but should rather suggest experiments to
determine whether similar mistakes will be unconsciously
repeated. By a little care and patience parents would
soon be able to arrive very nearly at certainty ; and the
complete application of Holmgren's test would only
be called for in cases where the existence of some
degree of defect had been rendered obvious in the more
simple way. In every such instance, of course, it would
be desirable to apply the test completely at as early an

age as the intelligence would allow ; so as to ascertain the precise degree and character of the defect, as well as the mere fact of its existence.

The different elements which compose white light are not only different in the sensations which they excite, but also in the degree in which they stimulate the retina. It is a matter of ordinary experience that some colours are more "trying" to the eyes than others ; and those in which red, yellow, or light green predominate are the most generally condemned. On this point Dr. Boehm of Berlin obtained some interesting evidence from the women who were employed, at a large establishment in that city, in embroidering ecclesiastical vestments ; and he was told that, while they could work upon a ground of blue satin until their fingers and heads were weary, they were soon compelled to lay aside work upon light green, on account of the distress which it occasioned to their eyes. On the whole, it may probably be said that the light of long wave-lengths, or of wave-lengths broken and disturbed by the interference of others, is more dis-tressing than that of shorter or more regular vibrations ; and it is well known that red, which has the longest wave-lengths, both impresses the retina more powerfully, and exhausts it more completely, than any other colour. Many toys have been founded upon the fact that a sort of temporary colour-blindness may be thus produced ; and the resulting phenomena have lately been pressed into the service of trade advertising. If we look steadily for a time at a red spot, seen under bright illumination, and then turn the eyes to a white surface, we shall first

see a similar red spot reproduced there from the per-
sistence of the original impression, and then, as this
fades away, it will be replaced by another spot of the
complementary colour, due to the circumstance that the
portion of the retina on which the red image fell has
become so far exhausted that it is for the time unable to
perceive the red element in the white light which falls
upon it, but sees only the bluish green produced by the
admixture of the still visible green and violet.

The properties of the different colours, and the different
ways and degrees in which the eyes are affected by
them, are matters not without some practical bearing
upon the furnishing and decoration of rooms, as well as
upon the choice of the means of obtaining artificial
light; and in the eleventh chapter it will be necessary
to return in some measure to the subject.

CHAPTER X.

THE number of blind persons in every civilised community is exceedingly large ; and, of the total number of the blind, the loss of sight dates, in a very considerable proportion of cases, from the first few days or weeks of life, or, as it is often erroneously said, from birth. I am not aware of the existence of any trustworthy statistics which show the precise proportion which those who become blind in early infancy bear to the rest of their fellow-sufferers, but common experience, or a visit to any one of the institutions for the education of blind children, will show that it must be large. The most recent statistics of blindness with which I am acquainted are derived from the census of the State of Massachusetts in 1875 ; and there, in a total population of 1,651,912, there were 2,806 blind persons, or 1 in 588. If we assume that the same proportion will apply to this country, there must be over 52,000 blind persons in the United Kingdom.

Of the persons who are commonly said to have been blind from birth, the enormous majority, probably at least ninety-nine out of every hundred, would bring with them into the world eyes as good and useful as those of their neighbours. The causes of infantile blindness are more frequently to be found in careless-ness and ignorance than in all possible injuries and diseases put together; and the carelessness and the ignorance are displayed, most frequently, in the neg-lect of proper precautions about light, cleanliness, and temperature.

It has been known from the earliest times, and has been abundantly confirmed by recent experience, that exposure to intense or dazzling light may not only pro-duce temporary or permanent diminution of the sensi-tiveness of the retina, but that it may also partially or completely destroy the power of vision. The eyes of infants are at least as sensitive to light as those of adults, possibly even more sensitive; and they are far less protected against it. From the imperfect deve-lopment of the bones of the infant skull the eyes are placed, so to speak, on the surface instead of being in hollows; the eyebrows and eyelashes are short, thin, and pale, the eyelids are almost transparent, and the irides are imperfectly furnished with opaque pigment. In the first weeks of life, moreover, infants are unable to shelter themselves from dazzling light by changing the position of the head. The importance of these several conditions should be more considered by parents and nurses, with reference to the regulation of the light which is admitted to the cradle, than seems usually to be the case. We find only too often that an infant is

placed close to a window in the full light of day, and even
with the sun shining directly upon its face. This should
never be permitted, although there is no reason for
falling into the opposite errors of covering the face so
as to impede the access of fresh air, or of keeping
the room so dark as to render the eyes preternaturally
sensitive.

Next after the precautions which are essential to the
maintenance of life, the cleansing of the eyes of the
new-born should receive early and careful attention.
This cleansing should be finished before any attempt
is made to wash the head or body ; and it should be
performed with water of a gentle warmth, and with
pieces of soft linen set apart for the purpose, or with a
small fine soft sponge first carefully scalded and purified.
The water employed should be contained in a convenient
basin, into which no part of the infant should be dipped ;
and the not uncommon practice of placing the infant in
a bath, and then of washing the eyes, first, indeed, in
point of time, but with the water of the bath in which
the body is already immersed, cannot be too strongly condemned. The eyes should be completely cleansed and
dried before any other cleansing is commenced, and they
should not again be touched until the next period of
washing, which should be performed in the same manner
as at first, so as to prevent the possibility of injury from
soap or any other irritant.

The prompt washing of the eyes has the important
advantage of speedily removing any hurtful matters
which may have come in contact with them; so that

the smarting or discomfort thus occasioned, as well as the possible danger, if not entirely relieved or obviated, will at least be greatly diminished. The best liquid for the purpose is simple warm water; and by preference river or rain water, since the spring water of some localities contains mineral substances, the presence of which might not be a matter of indifference to such delicate organs. The ingredients sometimes added to water by nurses, such as white of egg, infusion of marsh-mallows, milk, and so forth, are at the best only harm-less; and pure warm water should always be preferred.

Impure air may affect the eyes of infants injuriously. The foulness due to overcrowding, or to the presence of dirty clothes and such matters, is even worse than smoke or dust; and the vapour produced by the washing of clothing may often be loaded with particles of noxious organic matter, which may thus be conveyed into the eyes as readily as into the lungs.

Exposure to cold, either by draught upon the face or by the subjection of the whole body to a sudden change of temperature, is among the injurious influences which may injure the eyes; and a chill of the whole body may be occasioned by putting on damp or cold napkins or garments.

But the greatest danger to which the eyes of infants are exposed is from the inflammatory disease called purulent ophthalmia, which may occur notwithstanding the greatest care. The inflammation begins, generally speaking, between the second and the fifth day, or it may be somewhat longer delayed. It shows itself by

redness and swelling of the eyelids, and by the formation
of a thick yellowish discharge, which at first resembles
mucus, but soon assumes the characters of matter. At
first this discharge is scanty, and glues the eyelids together as it dries, but it quickly becomes more abundant,
and may then escape freely. As soon as swelling and
discharge are observed, no time should be lost in obtaining proper medical assistance. The inflammation
may subside, and the eyes escape, even without treatment; but in many cases irreparable mischief is done
even in a few hours, and no unskilled person can distinguish the most trivial from the most dangerous case.
It must never be forgotten that this disease is the
chief cause of blindness in infancy; and that a short
period of neglect or of unskilful management may lead
to partial or complete destruction of the cornea, escape
of some of the contents of the eyeball, and subsequent
wasting or other deformity. Every ophthalmic surgeon
can speak from only too much experience of the eyes
which are lost in consequence of this affection, especially
among the poor, simply from neglect or delay to obtain
proper advice. Scarcely a week passes in which I do
not see at the hospital some poor little baby in whom
purulent ophthalmia has been left for three or four
days or longer to the devices of grandmothers or nurses,
and whose sight, in one or both eyes, is in consequence
hopelessly destroyed. I will not go so far as to affirm
that sight would never be lost under skilful medical
treatment, but the cases in which it would even suffer
would be extremely few in number; and I say without

hesitation that I have never myself seen a single instance
in which an eye has been permanently damaged, if it were
taken under proper care before irremediable mischief was
already done.

Until medical assistance can be obtained, the mother
should turn a deaf ear to all suggestions of domestic
remedies, and should content herself by keeping the infant in a comfortably warmed (65° to 70° Fahrenheit) and
moderately darkened room, and by the preservation of
the most perfect cleanliness. The window-blinds should
not be red or yellow, and should be double if there is
direct exposure to the sun. The eyes should be cleansed
and bathed with lukewarm water as often as any considerable quantity of fresh secretion is formed, and this,
in severe cases, will be perhaps every half-hour. The
secretion is itself actively irritating, so that it can hardly
be removed too carefully. For this purpose, the lower
lid should be gently drawn down towards the cheek with
the forefinger of one hand, while from the other a slender
stream of water is allowed to trickle upon the inside of
the lid, and is caught either by a sponge or a small cup
held against the cheek by another person. If the lids
are much swollen, or if the child is very sensitive to
light, it may be necessary that the upper lid should
also be raised by one finger of the second person, who
should tenderly draw it upwards towards the eyebrow,
and must be most careful to exert no pressure upon the
eyeball. In order to avoid any sudden movements of
the head, the spare hand and fingers of the assistant
may steady it on either side. When all discharge has

been washed away, the lids must be dried without friction, by gentle pressure with an old handkerchief or with some other soft and absorbent material.

Those who are in charge of the child must remember that the matter from the lids is extremely contagious, and that the smallest portion of it, introduced into a healthy eye by the medium of a finger or handkerchief, would be prone to excite violent and dangerous inflammation. They must therefore be exceedingly careful not to touch their own eyes whilst engaged in the cleansing process described above, to wash their hands immediately after it, and to prevent any careless use by others of sponges, linen, or towels, which may have been soiled by the discharge. The more severe the inflammation the greater will be the danger of infection; and therefore, both in the very rare cases in which only one eye is attacked, and in the more common ones in which one eye is far worse than its fellow, an endeavour should be made to prevent accidental inoculation of the sound or of the better one by the flowing of the discharge over the bridge of the nose. Where the difference between the two is marked, each one should have its separate basin, sponges and linen.

The only local application which should be ventured upon, before medical assistance is obtained, is a soft compress of folded linen, moistened in cold water, laid over the closed lids, and changed as often as it becomes warm. There can be no doubt that this is always comforting, and often beneficial. The medical treatment consists chiefly in the use of astringents; but any details with regard to it would be wholly foreign to the scope of the present treatise.

When the period of infancy is passed, and as soon as children begin to employ their eyes intelligently about surrounding objects, the time has arrived when the character of the visual function should be made the subject of observation. It is well known that the differences which exist among adults, in respect to the distance, the acuteness, and the duration of vision are exceedingly great. One person, who reads the finest print near to the eyes, will scarcely recognise friends when they are two or three yards away ; while another, who can see the hands of a turret-clock half a mile off, may require spectacles in order to read at all. One person can read, write, or otherwise apply the eyes to near objects, for fifteen hours or more daily without inconvenience ; while another cannot work in a similar manner for as much as a single hour. There are even many, in whom the eyes present no trace of disease, but who have not acute vision at any distance. In some of the preceding chapters the causes of such differences have been fully explained ; and they have been traced to differences in the original formation of the organs of vision, by which these are more or less calculated for continued effort. Apart from these differences, it must be admitted that the powers of the eyes, like those of the other senses, are capable of being improved by judicious use and cultivation, and of being impaired by the operation of various adverse influences. A delicately organised system may break down under a kind or an amount of labour which would serve to call forth and develop the strength of the strong. Parents are too much accustomed to think of and to treat children as if they were all born with eyes of similar formation and endurance ; and this error is productive of many evils, which begin to show themselves about the time when systematic instruction is commenced. Such evils are especially produced in cases of ordinary myopia, hypermetropia, or astigmatism ; for these departures from the natural shape and proportions of the eyeball do not declare themselves, to the perceptions of un-skilled people, with the same readiness as the varieties of imperfect vision which depend upon cloudiness of any of the refracting media, upon absence of pigment, or upon other conspicuous defect. When any of the latter are present, the behaviour of the children with reference to small objects is generally sufficiently declared to excite, even in the minds of the unobservant, some suspicion of the truth.

In children with eyes of the hypermetropic formation, squinting is very frequently produced, and will be of the convergent or in-turned variety. It is especially prone to occur when there is any other cause of imperfection of sight, as from slight cloudiness of the cornea left behind by inflammation ; and the tendency to deviation inwards may be much promoted by very small play-things, which are necessarily brought close to the face. The tendency is still more promoted, if such things are played with in dim light ; or if the child is taken but little into the open air, where he has a wider field of vision, or if, when there, he is not induced to look about him at distant objects. The cutting out of small figures from pictures, and the putting together of fine dissected puzzles, are in the same way disadvantageous; and it is better to give hypermetropic children amusements which do not call upon their eyes for any great amount of accommodation. Outdoor pursuits and exercises, skipping, building with wooden bricks, games with balls, and the direction of the attention to natural objects, are all especially to be recommended.

The majority of the short-sighted are to be found, without doubt, among the dwellers in towns and among the more educated classes. This depends upon the fact that the occupations, especially in childhood, exert a great influence upon the development of the higher grades of the affection. The lesser grades constantly pass undetected by common observation, and the eyes in which they might be discovered are supposed to be natural. In circumstances which early promote the application of the eyes to near objects we find a comparatively large number of children who, even at the age when they begin to learn to read and write, are already more or less short-sighted. It would appear that their eyes are not only naturally somewhat larger than normal ones, but that they are also inclosed in weaker and more distensible tunics, so that, under the same conditions which promote the occurrence of squint in the hypermetropic, they undergo a gradual stretching at the posterior pole, and a consequent elongation in the direction of the axis. There is

every reason to believe that the deviation of such
eyes from the normal form would not occur, or would
not occur in so great a degree, if their gaze were not
habitually directed to near objects ; and also, as the firmness of the tunics increases with the increasing strength
of the body, that mere delay in the application of the
eyes to near work would be decidedly advantageous.
We must therefore strongly condemn the practice of
teaching children to read and write at too early an
age, as at five or even four years. The premature
acquisition of these merely mechanical powers is of
no real advantage ; especially when they are only to
be obtained at the cost of some imperfection of development, either in the body as a whole or in some
single organ.

There is yet another matter to which reference may
be briefly made, although, in strictness, it does not
belong to this part of the subject. The proper use
of the eyes is, as has already been set forth, a matter
of education ; and, although this education is generally
an unconscious process, it is nevertheless one which may
be much promoted by judicious interference. When
children are noticing, or playing with, their toys, or
pictures, or any common objects around them, it is
very desirable to take some trouble in order to guide
them towards the acquirement of habits of careful visual
observation. If, for example, a child has a picture of
a dog, it is well to direct his attention successively, by
questions or remarks, to all parts of the drawing ; to
the head, the tail, the feet, the eyes, the ears, and so

forth, in order to accustom him, when looking at a
picture or an object, to take note of all its details.
Somewhat later, attention may be directed to the differences in size, colour, shape, and other particulars,
say between a cat and a dog ; and again, when the
child begins to think, he may be asked about the
length or the height of an object, or the distance of
one object from another, so as to promote the development of the muscular sense by which we judge of
distance and magnitude. In showing pictures, children
should not be suffered to pass hurriedly from one to
another, but should be induced to fix their attention
for a while upon the salient points of each, so as to
form a habit of minute and careful, rather than of
superficial observation. The power of mentally recalling the peculiarities of objects depends, at least
in a great degree, upon the way in which visual observation is habitually exercised ; and many persons in
other respects of good abilities would be far more
efficient in their respective callings, if they had been
taught to observe carefully, minutely, and correctly,
at the time of life when they were learning to see.
If two people witness the same incidents, or examine
the same thing, how great will often be the difference
between their attempts to recall and describe the peculiarities which have thus been brought under their
notice. Robert Houdin, in a well-known passage in
his autobiography, gives a highly interesting account of
the manner in which his powers of visual observation
were cultivated ; and, in a book which was the delight

of children fifty years ago, the *Evenings at Home* of
Mrs. Barbauld and Professor Aiken, there was a story
called "Eyes and no Eyes, or the Art of Seeing," which
deserves to live in literature, and which admirably enforced the lesson that it has been the object of this
paragraph to teach.

The time at which children are permitted to begin
close and sustained attention to ocular impressions exerts
a very decided influence upon their power of maintaining
visual effort at a later period. It is not at all uncommon
for the eyes, during the years of school life, to become
less acute, weaker, and more or less short-sighted. It
is very important, whether at home or at school, to see
that children in reading, and more especially in writing,
maintain a posture in which the head is not suffered to fall
too far forwards. With print or writing of the usual size,
the paper need never be nearer than twelve or at least
ten inches from a normal eye. If it is noticed that any
child brings his work habitually nearer than this, or that
he cannot decipher print at this distance, surgical advice
should be sought for him without delay. The approximation of the work may depend upon several causes ;
upon opacities or turbidities left behind by previous
inflammation, upon simple myopia or hypermetropia,
or upon absolute weakness of sight. The existence of
any of these conditions requires either medical treatment,
or especial care in the conduct of the education. In
myopia or hypermetropia, the use of properly selected
spectacles may be desirable or even necessary. In learning to write, it may be necessary to employ none but

N

large-hand copies, and to see that these are reproduced in the same magnitude. For this purpose it may be desirable to give double lines; and the use of a somewhat thicker pen-handle than usual may also be advisable. Children should never be permitted to use books with stinted margins, printed in small and closely compressed type, like too many of the hand-books which are supposed to be prepared expressly for their benefit. There are few schools in which the seats and desks are adapted to the sizes of the children, and in many they are all alike for children of widely different ages. Those for young children should be lower than others ; and all should be so proportioned as neither to bring the heads of the occupants too near the desk, nor to require irksome stooping of the body in order to place the eyes at the proper distance of twelve inches or so from their work. Too much bending of the neck will impede the return of blood from the head and from the eyes, as may be seen by observing the flushing of the face which occurs when the head is kept for some time in such an attitude.

Next, the children should never be suffered to read, to write, or even to draw, by an insufficient light. Nothing distresses the eyes more readily than failure in this requirement, and there is no other in which failure is more common. There are unfortunately many schools, from the elementary to the highest, which are so badly constructed as regards their windows that twilight commences in them quite early in a winter's afternoon, even when it does not exist all day. Whenever it is

not possible to effect alterations which will admit more
light, the master should strive to render the comparative
darkness as little as possible injurious to his scholars, by
such a re-arrangement of their work as may bring the
tasks which require the least eyesight to the darkest hours.
Where the light is always defective, the reading, writing,
and drawing should not be pursued for more than an
hour without interruption ; and short periods of other
work, or even intervals of complete repose, should be
made to intervene.

Most important of all, however, is the general principle that children should not be overburdened with
any tasks which call upon the eyes for close application.
" How many of my schoolfellows," writes Professor Arlt,
"have had occasion to deplore the doctrine of our
teacher, 'that by writing we learn.'" A teacher who
considers the eyes of his pupils will discover some
better punishment than that of making them write out,
ten or twenty times, an imperfectly learned lesson.
The greater number of eyes will bear much exertion
without sustaining injury, when once adult age is attained ; but a large proportion of them, during childhood, are too delicate for the amount and kind of work
which is now commonly exacted from them. There has
lately appeared, in the *Times*, a correspondence upon
the possible effect of written impositions in spoiling the
handwriting ; but those who took part in the controversy
seemed to have no notion of the more important issues
which it might raise. The knowledge of the injurious
effects of certain kinds of schooling upon vision is not

a new acquisition ; for Beer wrote, more than sixty years
ago, " He who has taken the fruitless pains, as often as I
have done, to try and impress upon parents and friends,
in the most friendly manner and upon the most con-
vincing grounds, the mischievous effect upon the eyes
of growing children of the forcing-house system of the
present day, will still be disheartened to find his well-
intended counsel, based upon long experience, and often
repeated, either entirely neglected or listened to only by a
few. . . . .Because people hold the imperfectly-understood
principle that children should be constantly occupied,
there is at all hours of the day a master at hand. There
is reading, writing, language-learning, drawing, arith-
metic, embroidery, singing, piano- and guitar-playing
without end, until the persecuted victims are rendered
pale, weak, and sickly, and to such an extent short-
sighted or weak-sighted, that finally medical counsel must
be obtained. . . . .Of what avail is it to many charming
girls, many estimable women, that as children they were
regarded as prodigies, when the soundness of their eyes
and the acuteness of their vision have been sacrificed. . . . .
I have seen pictures, worked upon a tobacco-pouch in
the so-called pearl-stitch, which were scarcely inferior to
miniature-painting, and which I examined with much
pleasure until I remembered the eyes of the em-
broideress. I hope that the publicity of my opinion
may secure to some poor children the daily enjoyment
of even a single hour of fresh air and of the free movement of their limbs. . . . .In the present daily teaching
of children, the work most injurious to their sight is the

constant piano-practice from engraved notes ; since the
uniformity and the small size of these notes are calculated soon to fatigue and weaken the strongest eyes, as
any one may ascertain by experiment." When framing
the last sentence, Beer was probably recollecting some
cases of astigmatism which he had seen in the course of
his practice.   At the time when he wrote the passages
quoted above, the nature of astigmatism was not understood ; and it is manifest that this condition renders the
effort of reading music, which consists chiefly of the
horizontal stave and the vertical lines of the notes,
especially fatiguing and injurious.   In commenting upon
the above passages, Professor Arlt writes : " If the
illustrious Beer were now with us, he would not fail to
call attention to the injurious print of many books, as
the stereotyped editions of Latin, Greek, and German
classics, the pocket dictionaries, and the small maps,
which require a magnifying lens to render the names of
places discernible.   Parents and teachers should be very
careful that such books and maps are not used by the
children under their charge.   The number of those who,
in consequence of these books, have suffered in the
extent, duration, and clearness of their vision, is not
inconsiderable ; and I remember that I myself, when I
had completed my school education, was no longer able
to see a mountain an hour's journey distant, and which,
in my thirteenth year, I had seen from the same place
with perfect distinctness."

In the choice of a profession for children, the capa-
bilities of their eyes should never be left out of account.

The state of a young man, whose eyes refuse to perform his accustomed work, may be even more painful
than if he were blind ; and we should find fewer persons
in this condition if more care were taken to consider
the powers of the eyes before deciding upon an occupation. Eyes which within a few years would fail an engraver, a goldsmith, or a watchmaker, would last their
possessor his lifetime if he were an agriculturist, a gar-
dener, or employed in many other callings. He who
has sound and normal eyes may choose his occupation
without reference to them ; but he who is short-sighted,
or weak-sighted, or whose eyes are inclined to be inflamed, must endeavour fully to realise the claims
which an otherwise desirable calling will make upon
his sight ; and to understand the different ways in
which this 'or that kind of work may be injurious
to him.

It may perhaps be laid down as a general principle
that a child who is simply short-sighted, and who can
employ his eyes continuously, and with clear vision,
upon small objects, such as very fine print, so long as it
is near enough, may undertake work which requires accurate and continued seeing. Experience teaches that
merely short-sighted eyes, when the short sight has not
reached a very high degree, say not more than four
dioptrics, in early life, will bear without injury very fine
and continuous work. . . . . In the higher degrees of short-
sight, however, it is undesirable to engage in any occupation in which the vision must be directed by turns
to near and to distant objects, since the latter require

the use of concave lenses, which will increase the strain thrown upon the accommodation by the former.

Children who are the subjects of weak sight or hypermetropia, and who either cannot see near and small objects clearly, or cannot see them for long together, or only by the aid of convex glasses, should be dissuaded from engaging in occupations which will demand from them the application of the eyes to uniform work upon fine or small objects. The hypermetropic can, indeed, be greatly assisted by glasses, but these are not available in all pursuits.

Children who have often suffered from any of the various forms of inflammation of the eyes, which are incidental to early life, especially if they show any tendency towards relapse, or if they are still prone to irritation of the margins of the lids, should never be allowed to undertake any kind of work in which they will be exposed to dust, particularly woollen dust, to smoke, or to excessive perspiration from fire or heat.

Even when the eyes are of natural formation and acuteness, it would be improper to forget how much the power of sustained visual effort is dependent upon the general vigour of the muscular system. Girls of feeble frames and late development should avoid, on this account, the more sedentary forms of industry ; and should rather find employment in work that is comparatively coarse than in sewing, embroidery, or the like. The caution herein contained applies, also, in a still greater degree, when the eyes have been weak or inflamed during childhood.

In addition to the foregoing general principles, the whole education of children with delicate eyes should be regulated with some reference to their delicacy. For those who attend a day school, the distance and manner of the journey, and the protection to be afforded upon the way, require careful consideration, since various forms of inflammation of the eyes are caused, or at least promoted, by exposure to wet or to vicissitudes of weather. In all day schools there should be arrangements to allow of the removal of wet or damp clothing, and especially of wet or damp boots or shoes, before the children are suffered to settle down to their tasks. The atmosphere of schoolrooms, and the due supply of fresh air to them, are matters which probably will not be regarded until school-boards and school managers have no political or polemical questions left to dispute about.

With regard to the actual conduct of the teaching, it must be remembered that there is no reasonable doubt of the injurious influence of premature exertion of the brain in retarding the development of the body, the eyes, of course, included ; and I myself entertain none that such premature exertion is at least equally injurious to the mental faculties themselves. Many years ago, I wrote an essay upon " The Artificial Production of Stupidity in Schools," which had for its purpose to show the manner in which the proceedings of teachers may defeat their supposed objects ; and this essay has now been so often reprinted, in this and other countries, that I would fain hope it may have induced some few teachers to reconsider their ways. For the present purpose, it is sufficient to observe that any excess of school work implies, almost of necessity, an undue application of vision to near objects ; and that hence, when the eyes are either weakly or diseased, such excess should be strictly prohibited.

It is very worthy of note that, in the experience of ophthalmic surgeons, it is exceptional to meet with a child suffering from defective vision who has not, before the defect was discovered, been repeatedly and systematically punished by teachers or schoolmasters for supposed obstinacy or stupidity. The very reverse of this practice is that which ought to obtain ; and apparent obstinacy or stupidity should lead, from the first, to the question, "Can he see perfectly?" Children have an indefeasible claim upon their elders for friendly and considerate treatment. If they are harshly or unjustly dealt with, punished for errors which they cannot avoid, or forced to undertake tasks, either mental or bodily, which are beyond their powers, they will suffer either in mind or body or in both. Unfortunately, the work of teaching seems to exert a destructive influence upon the imagination, using that word in its true scientific sense ; and the average schoolmaster has often done an amount of wrong which can hardly be repaired, before the surgeon has any opportunity of interposing to put the saddle upon the right horse, and to assign the palm of stupidity to the pedagogue instead of to the pupil.

It would be improper to leave the subject of the care of the eyes in childhood without some reference to the frequently contagious character of the forms of

superficial ophthalmia from which children are especially
liable to suffer, and which are often widely diffused
through the agency of schools. There is a contagious
inflammation affecting the whole surface of the conjunctiva, and secondarily the cornea, which has again
and again prevailed as an epidemic in workhouse
schools, more especially in the very large ones which are
attached to city unions, and which has destroyed the
sight of hundreds of children. There is also a contagious
form of inflammation of the lid margins, and of the
roots of the eyelashes, which does an immense amount
of mischief, and which is common among poor children
who live in large towns. The sufferers from this affection
come in numbers to the ophthalmic departments of the
London hospitals ; but the great difficulty in the way of
curing them is the perpetual worry in which their parents
are kept by the officers of the School Board. That
august body has so much respect for the medical profession as to attribute to its members the gift of prophecy ;
for it has actually put forth a blank form of certificate
of illness, in which the doctor who signs it is asked to
state, not only that the child referred to is then unfit to
attend school, but also how long it will be before this
unfitness will terminate. With so high an opinion of
medical knowledge as this demand indicates, the School
Board might surely allow its agents to receive a hospital
letter, dated in such a manner as to show actual and
regular attendance, as a sufficient excuse for temporary
absence from the ministrations of the teacher ; for it is
hardly fair to call upon hospital medical officers, whose

time is generally very fully occupied about more important matters, to write or sign certificates by the dozen in proof of the existence of conditions which are sufficiently evident to the meanest capacity. To call upon a child with sore eyes to go to school is in most cases an act of almost inconceivable folly as regards the child itself, and of very manifest cruelty to its unfortunate classmates.

CHAPTER XI.

THE eyes may suffer injury in adult age from a great variety of causes ; from defective or excessive illumination ; from excessive application ; from unclean or impure air ; from exposure to cold ; from mechanical or chemical injury ; from mental conditions, and from many-so-called pleasures ; from unnatural conditions of the general system, which either occasion determination of blood to the eyes and to the head, or which depend upon abnormal states of the blood, by which the strength of the whole body, and with it that of the eyes, is reduced ; and finally from the want or the misuse of spectacles. The general principles which should be borne in mind, in order that the evils hence arising may be avoided, are of importance to every one who wishes to preserve his eyes in a state of health and of efficiency.

With reference to *light and illumination*, we must carefully distinguish the natural or white light which comes to us from the sun, from the artificial and coloured light which we obtain by the burning of various substances. The natural light is as congenial and necessary to the eye as food to the digestive organs ; but, just as these organs may be so enfeebled by long abstinence that they can only tolerate food in very small quantities, so the eyes, by the prolonged exclusion of daylight, may be brought into a state in which even a very moderate amount of light serves to irritate them, and a strong light is absolutely unendurable. In this way it is possible for the prolonged exclusion of daylight to prove highly injurious, even such exclusion as is produced when the eyes are bound up for a long period, or from the continued wearing of dark-coloured glasses. Some cases have been recorded, notably that of Dr. Harley, which show that the amount of light necessary to afford vision depends rather upon the amount to which the eyes have become habituated than upon the actual quantity. We all know how soon vision is recovered after coming into a comparatively dark room ; and Dr. Harley and others, after having been kept for a time in what would be described in common parlance as complete darkness, have been able to distinguish the objects around them. Persons imprisoned in dungeons, to which no light found access but by some single fissure, have after a while been able to see in this darkness, and to distinguish the mice which were attracted thither in search of crumbs. In such cases, the return to daylight has at first been very painful ; but, if cautiously and gradually managed, has seldom been productive of any permanent injury.

On the other hand, vision has often been enfeebled or destroyed by exposure to a dazzling light, either suddenly on leaving comparative darkness, or even without the aid of the contrast thus afforded. Every one knows the discomfort which is experienced from a sudden increase of light ; as when we come out of a dark place into one which is brilliantly illuminated, or when a flash from some reflecting surface is suddenly allowed to fall upon the eye. Such discomfort should be regarded as a warning of the dangerous character of the circumstances which call it forth. Many people have expiated, by impairment or loss of sight, the indiscretion of looking at the sun, especially when an eclipse has been observed through a piece of insufficiently obscured glass. After looking at the sun there remains, in less disastrous cases, an appearance as of a dark disc or cloud, which may assume a fiery or a violet tint when the lids are closed. This appearance may continue for a variable time, hours, days, or weeks ; and may even leave a permanent dark spot in the middle of the field, in such a position that it will interfere more or less with the vision of minute objects. In other instances, the disturbance of sight is not limited to a single spot, but extends, either gradually or immediately, over the whole of the retina ; and the most prompt treatment has sometimes barely sufficed to prevent the occurrence of complete blindness. Professor

Arlt records that he saw three cases of this description after the solar eclipse of 1851.

Temporary or permanent loss of vision has been occasionally produced by the reflection of the solar rays into an eye from a mirror or other like surface. Mischievous children have sometimes produced blindness in this manner by playing with a mirror.

A long continued gaze upon the full moon is said to have produced marked weakness or distress to very sensitive eyes, and it is easy to conceive that exposure to strong artificial light, as that of fireworks, Bengal lights, the electric light, white hot metal, and so forth, may prove injurious. Beer has recorded instances in which exposure to bright sunlight on awaking in the morning has produced troublesome inflammation ; and everybody knows how much temporary distress may be occasioned to the eyes in this manner, either by sleeping in a bed facing the morning sun, with the windows of the room insufficiently covered by curtains, or when these curtains are suddenly thrown open by a servant. Persons who use a night light should be careful so to place and to screen it that its rays do not fall directly upon the eyes ; but there can be no doubt that such an appliance is much better omitted than used ; and that perfect darkness is eminently conducive to the attainment of sound and refreshing sleep.

The solar light when reflected from white surfaces has often been injurious. The inhabitants of polar countries have found by experience the necessity of

protecting themselves against the glare from snow;
and surfaces of sand, of chalk, or even of water, have
often done harm to voyagers or travellers by the brightness of the reflection which they have afforded. Sailors,
who are inhabitants of temperate climates, and who
have been much exposed to glare in the tropics, have
not seldom been affected by night-blindness; a condition which depends upon over-stimulation of the
retina, by which it has been rendered insensitive to
anything less than the illumination of full daylight.
All travellers going to the tropics, or into countries
where there is likely to be snow upon the ground
together with sunshine, should provide themselves with
dark-coloured protecting glasses, or with some similar
protective contrivances.

It follows from the foregoing considerations that the
manner of admitting light into the rooms of a dwelling
house is by no means a matter of indifference to the
inhabitants. Of artificial light I shall have to speak
hereafter; but we may often find rooms in which there
is far too much daylight. It should always be remembered that the position of the eyebrows, and the general
structure and arrangement of the eyes and their appen-
dages, are calculated chiefly to protect them from light
coming down from above, and that they are comparatively defenceless against that which comes up from
below. On this account, very low windows are rather
to be avoided; or, if used, they should be fitted with
blinds made to draw up rather than down; and the
floors should not be covered with very bright coloured

materials, or with any which possess reflecting surfaces. The blinds, too, by which the admitted light is tempered, should be of a suitable colour ; neither white nor white striped with red ; but of a blue or grey tint, and of sufficient thickness to be really effectual for the purpose for which they are designed.

Those whose occupations or manner of life call upon them for regular use of the eyes upon small matters, as in reading, writing, keeping accounts, embroidering, or any similar pursuits, have not only, like all other persons, to consider the question of light in general, but also the kind of illumination, whether natural or artificial, by which their work is done ; and to consider this the more carefully, the finer the work itself may be, the less it is diversified by other pursuits, and the shorter the periods of intermission. Under defective illumination all eyes are prone to suffer in acuteness of vision and in the power of sustained effort ; and the eyes of young persons are likely to suffer also in visual distance.

The conditions of illumination which are most pre-judicial to the eyes compelled to work in them are chiefly these : when the light is too feeble, or insufficient ; when it is too strong, so as to be dazzling ; unsteady, some-times strong and sometimes weak ; irregularly dis-tributed, as when broken by shadows ; of bad quality, as compared with white daylight ; and, lastly, when it falls upon the eyes or the work in a wrong direction.

Since all the foregoing forms of defect are most liable to be incidental to the use of artificial light, such as is

O

afforded by candles, lamps, or gas, it follows that those
who have no alternative but to work by such artificial
light should be especially careful with regard to its
quality, direction, and general management. It has
been mentioned in an earlier chapter that the solar light
is a compound of different colours, which can be
separated by passage through a prism ; and which are
ultimately resolvable into the three primaries, red,
green, and violet. These three, like the seven of the
spectrum, can be reunited so as to produce white light;
and in the solar light the three primary colours are
combined in the proportion of five parts of red, three
of green, and eight of violet. When this proportion is
changed, as it generally is in artificial light, the light is
no longer pure, or white, but is coloured ; and the colour
of that kind of primary light which is present in excess
is of course that which preponderates. In common
candle or lamplight, the red rays are in excess ; so that
the light itself is more or less reddish or orange. This
peculiarity of artificial light is productive of many
important consequences. The different colours produced
by dyes or pigments assume quite a different appearance,
when seen by lamplight, to that which they have by
daylight ; green appears yellowish ; blue, greenish, more
or less according to the greater or less quantity of violet
in the artificial light employed ; dark-blue, purple-red,
and orange appear much brighter, in proportion to the
excess of red rays, and so on. In bright moonlight,
if we look at a surface, say of the paved causeway of a
street, which is partly shaded from the moon, but lighted

everywhere by gas, we shall see that the part lighted only by gas will appear reddish, by contrast with the adjoining part which receives the moonbeams also, and which appears of its natural greyish tint. It is a pretty experiment to admit into a room with white walls the light from a white cloud, and to hold up a staff in such a manner as to cast a shadow upon the wall. This shadow will appear simply dark; but, if we bring in a lighted candle, and throw a second shadow from its flame by the side of the first one, then the first shadow will appear yellow, and the second will appear blue. The reason of this is that the first shadow, formed by the interception of daylight, will be lighted up by the light of the candle, in which yellow preponderates ; while the shadow formed by the interception of the candle light will be illuminated by the daylight, and the two being near together, produce an effect of contrast which causes the second to appear blue by comparison. If we surround the candle by a blue chimney glass, then both shadows will be alike, simply dark, from whence we may infer that both the sources of light contain the same colours, but in different proportions ; and that the blue glass, by absorbing red and green rays from the flame, restores something like the natural or solar proportion between its colours. It follows from these conditions that working with colours by artificial light is either impossible or highly fatiguing, especially with dark colours. The dark colours absorb too much of the light which they receive, and return too little, for them to be sufficiently illuminated from artificial sources.

Whenever possible, dark surfaces should be worked upon only by daylight, and lighter surfaces should be reserved for the hours of artificial light; but embroidery with and upon colours, which is sufficiently trying even by daylight, should never be attempted by artificial light at all.

Precisely as the powers of all other organs are exhausted by exertion, and require to be renewed by rest, so the eyes become dulled by the prolonged incidence of light, and require to be renewed in like manner, more especially if the light is not of the natural colour and quality. The necessary rest is afforded, to a great extent, by the mode in which the two eyes work together, it seeming that they see actively and passively by turns, each coming to the support of its fellow in time to save it from exhaustion. An observer who is placed in an astronomical chair, which supports the head and neck without fatigue in an appropriate position, may fix the eyes upon a single star, and may keep it in view for a very long period. If, however, the star is looked at with one eye only, it will before long disappear from view; the reason being that the portion of the retina upon which the image has fallen has been rendered insensitive to it by exhaustion, and requires a period of rest. If the lids are closed, and the gaze turned away, the star may again be seen with the same eye after a few minutes; but it will be lost on the second occasion more quickly than on the first. When the light to which the eyes are exposed is of an unnatural colour, the power of the second eye to support the first is less declared;

and those who work much by artificial light, in which there is the usual relative preponderance of red and green rays, are sometimes found to become comparatively insensitive to these colours, so that, when they are in white light, or looking upon white surfaces, they see an appearance of points, lines, or clouds, which are dark blue, or dark reddish-blue, or almost black. The principle of these appearances has been explained in a previous chapter, and there is a familiar child's book in which this principle is applied to the production of so-called apparitions. The book contains pictures in bright colours, and after looking at these pictures steadily for a time, if the eyes are turned towards a white surface, the same pictures are seen in the air in colours complementary to those in the book. Assuming the picture to be red, the eyes, by looking at it for a time, are rendered insensitive to the red rays. They are then turned towards the white light, in which they are capable of being impressed by the violet and the green rays only. Hence they see the appearance of a bluish green or green figure, precisely resembling the red one at which they have been looking. Such experiments are harmless when neither too often repeated nor too long continued ; but they show how easily the sight might be permanently injured by the constant employment of the eyes under artificial conditions. Furthermore, it follows from the composition of artificial light that its illuminating power does not bear any definite proportion to its quantity. Where, for example, there is an excess of the red rays, which afford but a feeble illumination as measured

by their power of displaying objects, it is manifest that
there must be more of the light as a whole, to allow of
any occupation being carried on, than would be necessary
if it were white ; and hence that the eye may be irritated
and dulled, not only by the primary excess of red, but
also by the secondary excess which depends upon the
greater quantity of the light that is rendered necessary
by reason of its inferior quality.

Lastly, the heating qualities of light must by no
means be left out of account. It is well known that
all combustion which affords light affords heat also ;
although the relative proportions of light and heat differ
in flames of different descriptions. A gas flame, for
example, gives much more heat than a candle flame of
the same magnitude. It has been made known by accurate experiments that the violet rays are the least heating, the green next, and the red the most, in the relative
proportions of 56, 58, and 72 ; and it follows that artificial light, in relation to its illuminating properties, must
be warmer, the greater the proportion of red and green
rays, and especially of red rays, which it contains.
Besides this, every ray of light is accompanied by a certain amount of free heat, which is found chiefly at the
red end of the prismatic spectrum, outside the light rays
themselves. Near objects, at which we have to look for
a considerable time, seldom derive their light directly
from the solar ray, which before reaching them has
been reflected from innumerable objects, such as atmospheric particles, clouds, the earth, plants and animals,
the furniture of rooms, and so forth, by each of which

the attendant heat rays have been more or less absorbed, so that the ultimate illumination is accompanied by comparatively little heat. When, however, artificial light is used, its rays generally fall directly from their source upon the object, and their source is commonly near to the worker ; so that both the light and its accompanying heat are usually reflected from the object of vision directly upon and into the eyes. This heat acts especially upon the surfaces of the eyes and upon the lids, drying and irritating them, and powerfully predisposing to various inflammatory affections. In certain callings, such as that of an engraver on wood, in which the source of artificial light is required to be near the worker, it is a common practice to interpose between the lamp and the object a globular glass vessel full of water, from which two advantages are obtained. In the first place, much of the heat is absorbed by the water ; and, in the second place, the globe serves to focalise or condense the light upon the point which is being looked at. On this account, a smaller and less heating flame than would otherwise be required may be employed, and the disturbance of vision by scattered illumination is wholly prevented. Another common plan is to place a large sponge, soaked in water, on the table near the worker. By this, no economy of light is obtained ; but the evaporation from the sponge serves to keep the surrounding air cool and moist, notwithstanding any heat which may proceed from the lamp.

In order to improve the colour of the artificial light from a lamp, it has been a common practice to surround the flame with a blue chimney, or to wear blue glasses. The glass for this purpose is coloured by oxide of cobalt, and, while it allows nearly all the violet rays to pass through, it arrests a considerable proportion of the red and of the green. Although such glasses have the advantage of rendering the artificial light nearer the colour of the natural, they have also the disadvantage of diminishing the total quantity of light which falls upon the object, and hence of entailing all the ordinary consequences of insufficient illumination. The blue glass, it must be remembered, supplies nothing, but only intercepts. Its peculiarity is to be very transparent to violet rays, and more or less opaque to others. Hence, when it transmits light in which red and green preponderate, it will to some extent restore the balance by arresting the preponderating elements. Generally speaking, it arrests too much of them to transmit white light; and it must be obvious that, the blue being originally deficient, and the red and green partially arrested in transmission, the total amount of light is considerably reduced. Unless the original light be very intense, the use of blue media will finally resolve itself into the use of defective illumination.

The other chief disadvantages of artificial light are its unsteadiness, its variableness, and the faulty directions from which it often proceeds. All the older forms of artificial light are more or less unsteady, from the absence of any means of preserving an uniform proportion between the air-supply and the combustible. This unsteadiness is always objectionable, and, in its worst form, as in working by the light of a flickering candle, it may be highly distressing. Attempts were formerly made to diminish the effect of this unsteadiness by the use of two or more candles, so that the inequalities of illumination depending upon irregularities of the wicks might be less noticeable, while the total quantity of light was increased. For this purpose, the candles should be placed as near together as possible, and should not be distributed, or divided to the right and left, since the latter arrangements not only diminish the total quantity of light which falls upon the work, but also interfere with its equal distribution. Candles, however, are now so little used, except as auxiliaries to other means of lighting, that it is hardly necessary to dwell upon their disadvantages.

A faulty position of the source of artificial light either sends the rays too directly into the eyes, or else affords only an insufficient illumination of the object looked at. An ordinary candle or lamp should have its flame a few inches higher than the eyes, in order to utilise the shelter of the brows ; and should be somewhat to the left front of the worker. If placed higher than this, the illumination is commonly insufficient for close application. The central gas or oil chandelier, by which rooms may be conveniently lighted for meals or conversation, seldom affords enough light for any purpose which requires accurate vision ; and I have constantly to warn parents against allowing their children to prepare lessons for the next day in a sitting room which is only lighted in this manner. A flame on the same level

with the eyes, or a little below them, sends too much
light directly into the pupils, and too much heat upon
the surfaces of the eyeballs, and thus tends to produce
dryness, dazzling, and irritation. Even upon these
grounds alone, the otherwise objectionable practice of
reading in bed should be prohibited ; since both the
light and the book are almost of necessity placed too
low and too much on one side, and the eyes are brought
into a strained and unnatural position. Where the habit
cannot be relinquished, it is at least desirable to have a
lamp which is capable of being set at any desired height.
When the light is too much on one side, it affords to
one eye a comparatively feeble, and to the other a comparatively strong illumination: The tendency of this inequality is to produce unequal contraction of the pupils,
and to render the visual act exceedingly fatiguing.

Of late years, great attention has been given to the
means of obtaining artificial light, and many improvements have been effected. The electric light furnishes
the nearest approach to the actual solar ray which we
possess, since it contains almost the same proportion of
violet ; but it is rendered dazzling by the directness with
which it comes to us from its source, and by the consequent absence of those multitudinous reflections to which,
in the case of sunlight, reference has previously been made,
and by which both its colour and its brightness are much
modified. Moreover, for domestic purposes, the electric
light is not yet practically accessible ; and, even if it were,
we should still have to learn by experience the effects
of continued exposure to an illumination which is as

rich in the actinic or chemical rays as in those which are recognised by our senses in the form of light. It is by virtue
of these actinic rays that light holds its place as one of
the most powerful of the stimulants to vital action ; and
nature is careful to withdraw them from us for long
periods. They might prove to be as harmless as the
illuminating rays themselves ; but that they are so has
yet to be ascertained by direct observation. In the meanwhile, the best artificial light is that which most nearly
approaches diffused daylight in colour ; and this character belongs, I believe, to that which is furnished by
the various forms of burners invented and patented by
Mr. Silber. It may be said, generally, that the quality
of light will be proportionate to the completeness of the
combustion by which it is afforded ; and complete combustion is only to be secured by the precise adaptation
of the quantity of the combustible to the quantity of
atmospheric air which finds access to the flame. When
the combustible is in excess, part of it must remain
unconsumed, or only partially consumed, and the flame
will be dark, smoky, and ill-coloured. When the air
is in excess, its superabundance will lower the temperature of the flame, and will thus diminish chemical
activity, with the same ultimate result as in the former
case. In the original Argand burners, although they
were a great improvement on all which had preceded
them, the air supply was in excess, so that the flame was
unduly cooled ; and no precise regulation was attained
until the construction of the Silber Argand burner,
which, in the form adapted for gas, is shown in elevation

and in section in Figs. 46 and 47. It will be seen
that the air only obtains access to the flame through
a square opening of carefully calculated dimensions;
that the entering current is divided, and that part of
it is conveyed through the inner tube to the interior of
the flame near its summit. By this contrivance, any
volatilised particles of carbon which are being driven off
unconsumed are brought into contact with a fresh supply
of oxygen, and are completely burned. Another point
of difference from the old Argand is that the supply of

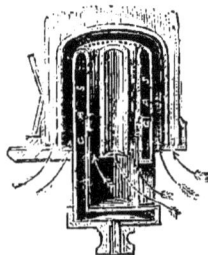

FIG. 46.　　　　　　　　　　FIG. 47.

air can only reach the flame by passing through metal
channels of considerable length, in which the air becomes
heated to such a degree that, even if superabundant, it
would exercise no refrigerating effect.

Coincidently with the researches of Mr. Silber, much
good work in the same direction has been done by
another inventor, Mr. Sugg, who also makes Argand gas-
burners of great excellence. As between the Silber and
the Sugg burners it would be difficult for the unaided
eye to assign the palm of superior merit ; but, according
to a photometric investigation made by Dr. Wallace, and

the results of which were embodied in a paper read by him before the Society of Arts, and published in the Journal of the Society for the 7th February, 1879, the Silber burners are decidedly, although not very greatly, the better of the two. For the same consumption of the same kind of gas, the light given by the best form of Silber Argand burner exceeds that given by the best form of Sugg Argand in the proportion of about twenty-five to twenty-two. It is only in gas-burners that the two inventors are competitors; and in burners for oil, to which the Silber principle is also very completely applied, I am not acquainted with any others which at all approach them in either the quantity or the quality of the resulting illumination.

The weak point common to all forms of artificial light, other than the electric, is their original deficiency in violet rays; and some time ago, with the kind assistance of Professor Barff, I commenced a series of experiments, which I was unfortunately not able to complete, for the purpose of ascertaining the relative merits of different kinds of burners and of different kinds of illuminating agents, with especial reference to this single question. The tests which may be used are two in number; the first being the examination and comparison of known colours by the artificial light, the second being its actual analysis by a prism, which separates its component elements and allows the quantity of each to be measured. When tested by the prism, all artificial light is poor in violet; but that afforded by the Silber burners contains more than any other which I examined. The Sugg burners had not been reached when the experiments

were interrupted ; but I found, as a general result, that
the amount of violet was proportionate to the amount of
illumination, and hence I entertain little doubt that
the Sugg burners are inferior to the Silber in both
respects to about the same degree. Professor Barff, in
his lectures at the Royal Academy, had already called
attention to the value of the Silber burners as sources
of an artificial light by which colours, even the most
delicate shades of yellow, could be distinguished apart
upon a white ground; and by which it was possible
to paint pictures that would bear the test of daylight
inspection. His statement has now been fully confirmed by the experience of artists; and the Silber
burners form an important part of the fittings of
many studios.

Whether the illuminating agent employed shall be
gas or oil, and whether the oil shall be colza or petroleum, are questions mainly of economy and convenience.
The Silber burners reduce the evils of gas to a minimum ;
inasmuch as these evils, such as the liberation of substances which are either injurious to human beings, or
destructive to book-bindings, pictures, and furniture, are
almost entirely due to imperfect combustion, and are
prevented in proportion as the combustion is complete.
I have no information with regard to the quantity of air
which is consumed, light for light, by gas as compared
with oil ; but it is abundantly sure that gas consumes
a great deal, and that any room in which it is burned
should be thoroughly ventilated. Notwithstanding the
greater trouble which it gives, I confess to a decided
preference for oil, and, among oils, to colza as compared

with petroleum; because the latter possesses physical properties by virtue of which it is almost impossible to prevent leakage. When there is no leakage, properly so called, there is exudation; the oil creeping in some mysterious way out of its reservoir and diffusing itself in a thin film over the surface of the lamp, which cannot be touched without greasing the fingers. From this source of annoyance the use of the less volatile and more viscid colza oil affords entire exemption.

Since the foregoing paragraphs were written, the *Observer* and other newspapers have given prominence to a passage, extracted from an inferior English medical journal, and purporting to contain a summary of a lecture on the action of light upon the eye, recently delivered in Paris by Dr. Bouchardat, Professor of Hygiene to the Faculty of Medicine. The summary was so meagre and imperfect, that no conclusions could be based upon it; but it distinctly put forth a recommendation, in the name of science, that we should return to the use of the tallow-candles of our ancestors, and this was sufficiently remarkable to induce me to procure the original lecture. In speaking of the light yielded by candles, by petroleum, by gas, and so forth, the venerable professor dwells at length upon many imperfections, the results of irregular or incomplete combustion, which have no essential relation to any of the agents mentioned, but which depend upon the faulty preparation of those agents, or upon remediable faults of detail in the construction of lamps or burners, or in the regulation of their air-supply. Beyond this, he does really fall

into the strange error of asserting that the most advantageous varieties of artificial light are those which
contain the largest amount of yellow in proportion to
their red and violet ; and he adds that the substances
which fulfil this condition are the several animal and
vegetable fats. It is manifest that he supposes yellow
to be one of the primary colours of the spectrum ; and
the want of acquaintance with modern researches which
he thus betrays is alone sufficient to cast some doubt
upon the value of his teaching. He believes, also, that '
the violet rays of the spectrum are injurious to the eyes ;
and this belief seems to rest upon a hypothesis in support
of which, as far as I am aware, no facts have ever been
adduced. The violet and ultra-violet rays produce the
phenomenon called " fluorescence " in the crystalline lens
and in the other media of the eye, as well as in most
transparent substances ; and the hypothesis is that this
fluorescence is a result of molecular vibrations which
must be injurious. Why they should be injurious is
not stated ; and there is not a particle of evidence that
they are so. Fluorescence is a faint luminosity which
proceeds for a time from various substances which have
been exposed to the violet rays, or to mixed light containing them ; and it may be described as a minor degree
of that "phosphorescence" which is displayed by the
sulphurets of calcium and of barium, and which has
recently been rendered familiar to the public by the
manufacture of clock-dials which are luminous at night.
Most physicists are agreed that some kind or degree of
molecular vibration is an universal condition throughout

nature; and why the molecular vibration of fluorescence should be assumed to be hurtful to the tissues of the eyeball is more than I can say. Dr. Bouchardat correctly maintains that the best artificial illumination is that which most closely resembles the light of the sun in temperate climates; but he forgets that this is only rich in yellow, when it has been robbed of a portion of its violet by atmospheric impurities. There can be no reasonable doubt, as already laid down, that the mixture of primaries contained in the solar spectrum is that which is most congenial to the eyes, and that every departure from this mixture produces more or less disturbance of the balance of vibrations to which the retina is best calculated to respond. Professor Barff maintains, I believe with entire accuracy, that the essential condition of harmony of colour in a picture, or in a stained window, is that the various tints should be in such proportions that they would produce white light if they were re-combined. The late period, at which the observations of Professor Bouchardat were made known to me, forbids me to examine them in greater detail; but it was plainly necessary to express my entire dissent from propositions which appear to me to be unsound in theory, and likely to be injurious in practice. I have no fear of violet rays in their due proportion; but I should have great fear of the consequences of continuous application of the eyes by the light, or rather by the semi-darkness, which would be afforded by the use of tallow-candles.

Whatever kind of light is employed, consideration

P

will be well bestowed upon the form of the lamp ; and,
the best form, for all purposes of table work, is that
shown in Fig. 48.   Lamps of this description, with
common Argand burners for colza oil, have been for
many years largely manufactured by a maker named
Stobwasser, of Berlin ; and they were imported into
this country, under the name of "Queen's Reading-
Lamps," by several tradesmen, some of whom attached

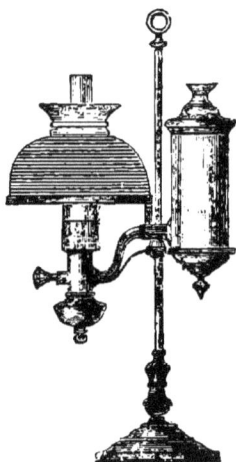

FIG. 48.

their own names to the lamps they sold, and charged for
them at a rate greatly in excess of their value.   Within
the last ten years I have seen these lamps, made by
the same maker, and identical in every respect, offered
for fifteen shillings at one shop and for two pounds ten
shillings at another, the two shops being scarcely five
hundred yards apart.   They are now made with Silber
burners, and the same construction is easily adapted to
gas.   Besides the convenience of being able to raise or

lower the burner at pleasure, they are recommended by their hemispherical shades, which are usually of glass, either white, or dark green with a white lining, the latter being for most uses to be preferred. All who work by lamplight should have the flame covered by a shade of this or of some similar description. The eyes are instinctively raised at every pause in the occupation ; and then, if there be no shade, they are raised to encounter an increased degree of illumination, because, whatever form of light is used, it is manifest that we receive less of it by reflection from the surface at which we are looking than if we look directly at the lamp itself. Without a shade, therefore, the eyes will be stimulated as often as they are raised, their pupils will be forced into sudden contraction, and they will be fatigued and worried during the moments when they might be at rest. With the shade, on the contrary, the eyes will be raised to comparative darkness, the pupils will be relaxed and will dilate, the retinæ will be soothed and rested, and effort will be resumed with increased ease and comfort. For some purposes, and when the eyes are irritable, I carry the benefits of the shade as far as possible, by having it made of metal, so that it may be perfectly opaque, and furnished with a sort of tongue in an upward direction, to stop all radiation of light from the exposed part of the chimney.

It is an evil incidental to perfect combustion that it is attended by the development of heat in proportion to its light ; and the Silber burners, either for gas or oil, are unquestionably very hot. Their influence in drying the air of a room may be prevented, of course, at least in a great measure, by proper ventilation, which necessarily implies a constant fresh supply from outside ; and, if this is not sufficient, the already-mentioned wet sponge upon the table will furnish an effectual remedy. The direct radiation of heat to the head and eyes, when engaged in reading or other sedentary work, is best prevented by the interposition of a transparent screen formed of some substance which is impervious, or nearly impervious, to the heat rays. The most easily procurable screen for this purpose is a flat cell, which may be six or eight inches square and half an inch thick, its sides formed of plate glass, and its cavity filled with a saturated solution of alum in water. Such a screen has been made for me by Mr. Ladd, of Beak Street, by whom it has been mounted in a simple wooden frame, so that it can be placed upon the table in any desired position, and raised or lowered to correspond with the height of the lamp-flame. As a matter of experiment, I lighted my Silber gas-burner one evening when the temperature of my room was 62°, and I placed two thermometers on stands, level with the flame, and each with its bulb six inches from the chimney of the burner. One thermometer was directly exposed to the flame, the other was separated from it by the interposition of the alum screen. At the end of an hour, the unscreened thermometer stood at 99°, and the screened one at 73°, so that the screen had intercepted 26° of heat ; and it had done this without itself becoming sufficiently warm to radiate in any appreciable manner. The solution of

alum should be run through a paper filter ; and the cell
should be washed out, and the loss by evaporation supplied, as often as may be required, which will perhaps be
once a week. The solution of alum may be coloured of
a bluish tint, if desired, by the addition of a few drops
of a solution of ammonio-sulphate of copper ; but it
must be remembered, as already explained, that in this
way we only take away red and green, and do not add
violet. The same effect may be produced by a slight
tint of cobalt blue in one of the sides of the cell ; but
it is necessary to have plenty of light to begin with
before attempting to improve its quality by either of
these expedients.

To sum up what has been said upon the subject, the
best illumination for all purposes of close work is that of
a Silber Argand burner, it matters not whether for gas
or oil, placed to the left front of the worker, a little above
the level of the eyes, fitted with a shade to reflect light
upon the table and to intercept it above, and with the
addition of an alum screen when the heat is objectionable. To this arrangement a wet sponge may be added
for those whose eyes are irritable, or whose eyelids are
prone to be inflamed.

Finally, with regard to all artificial illumination, those
who can divide their time as they please should endeavour to get through work which is trying to the eyes
in the daytime, and should reserve for evening that
which requires a less degree of visual effort. Reading
requires a greater degree of visual effort than copying,
and copying a greater degree than simple writing.

Furthermore, in the long continued use of artificial light, an occasional change or alternation of occupations is exceedingly desirable. After two or three hours of reading, for example, the eyes may be much rested by writing for half an hour ; and it is still better to break protracted work, of whatever kind, by short intervals of complete repose, during which the worker may walk about the room, and, if the eyes feel distressed, may bathe them with cool but not too cold water, to which a little toilet vinegar may generally be added with advantage.

Even to those who work only by day there is much to be said on the subject of illumination. In moderate daylight, the normally constructed and healthy eye is capable of great exertion, both as regards the fineness of the objects of vision and also as regards the duration of the visual effort. Notwithstanding this, those who would use their eyes continuously, and would at the same time preserve their sight uninjured, should attend, even by daylight, to certain prudential considerations. Thus, work should not be carried on in very strong or direct sunlight ; and persons who are engaged in dealing with polished surfaces should also avoid direct sunlight, which may be suddenly reflected into their eyes in a hurtful manner. The same applies to workers with the micro-scope, for whom the chief risk is that the sudden passage of a white cloud, for which their instrument has been adjusted, may throw into the observing eye or eyes the full glare of the sun. Several cases are on record in which great and permanent injury has been thus

occasioned ; and many of the sufferers have been medical
men or medical students, who were engaged in the dis-
section under the microscope of some white tissue, from
which the sudden increase of light was very strongly
reflected.

A more common fault, however, is working with an
insufficiency of illumination. The light should always
bear a due proportion to the fineness of the objects of
vision and to the colour of their surfaces. In reference
to this part of my subject, I cannot too strongly warn
my readers against the continuance of fine work by the
evening twilight. It is far more easy to apply the eyes
early in the morning, before the daylight is complete,
than to do so by the same degree of illumination in the
evening; because, in the latter case, they are already
more or less exhausted by the efforts of the day, and
moreover they have for some hours been accustomed to
the full daylight which is departing. When the fading
of light renders a lamp necessary, it is always desirable
to close the shutters, and to exclude whatever daylight
may yet remain; since this, although not sufficient to
work by, may yet be strong enough to weaken the artificial light. The fading daylight, if permitted to enter,
will fall chiefly upon the peripheral parts of the retina,
diminishing by comparison the brightness of the image
of the object of vision upon the yellow spot ; and ob-
servant persons soon discover that they see less clearly
by such mixed light than by either daylight or artificial
light singly. The best plan, however, is not to work
continuously over the transition from artificial light to

daylight, or *vice versâ*, either in the morning or in the evening, but to take an interval of repose between the two.

It is obviously undesirable to work behind green or red blinds, or to wear coloured spectacles when looking at very fine objects. By its passage through coloured material the light is necessarily altered in the proportions of its different elements; it is not only rendered weaker than the unchanged solar light, but also more irritating and less harmonious, and therefore less congenial to the eyes. The least objectionable coloured glasses are those of pale cobalt blue. Persons who are required to work in the face of direct or of directly reflected sunlight should never use the green glasses which are sometimes sold ; they are not much better than red or yellow. The best plan in such circumstances is to place a screen of ground glass, of a light grey or bluish white tint, before the window from which the light is derived.

For fine and trying occupations, such as drawing, embroidery, engraving, and the like, it is important that the light should fall upon the object in only one direction. Large panes of glass and large windows are always desirable for the eyes, and the glass should be free from flaws or irregularities. Broad window frames, sash bars, lattices and the like, interrupt the light too much to allow of the desirable uniformity being maintained ; even though they may not cast upon the work noticeable shadows.

People who read out of doors, and especially under the shade of trees, should be careful that they do not

read sometimes by direct sunlight, sometimes by the
comparative darkness of a shadow. Even walking up
and down in a room to read may be injurious, if the
successive turns produce frequent and sudden differences in the degree of the illumination. Reading in a
carriage is also highly undesirable ; not only on account
of the dazzling from inequalities of light, and from light
falling in upon the page from one or other side ; but
also, especially in a railway carriage, from the oscillations of the vehicle, which occasion a constant slight
variation of the distance of the page from the face, and
necessarily a corresponding variation in the distance for
which the eyes must be accommodated. It is well
known, as a matter of common experience, that the
eyes cannot work in comfort in a place where the light
comes at once from two opposite directions ; and the
reason of this is that the side light falls in too large a
proportion upon lateral parts of the retina, and stimu-
lates them in excess of the stimulation of the central
parts. Such lateral illumination is even more distressing if the object actually looked at with central vision
should at the same time be receiving only a comparatively feeble illumination. In this way, discomfort and
fatigue of the eyes may be occasioned by polished or
glittering objects above or on either side of the worker,
or in a room which has windows on two opposite sides.

It is likewise undesirable, in the use of artificial light,
to bring it between the work and the eyes ; or by daylight to turn the back to a window. A work-table
should be so placed, whenever possible, that the light

neither falls upon it directly from the front nor quite from the side ; but in an intermediate direction. For writing, whenever possible, the light should come from above and from the left front. If the face is opposite the window, the eyes often receive too much light, especially if this is reflected from light-coloured walls, or from bright clouds. Where such a position is unavoidable, the evils incidental to it may be diminished by darkening the lower panes of the window, either with suitable blue or grey hangings, or with ground glass in the manner already mentioned. Waxed blue paper may also be employed as a screen ; but it should be so placed as to allow the light to pass through it in an almost perpendicular direction, for which purpose its upper margin should be inclined towards the worker. When light comes directly from one side, it affects one eye more than the other, and hence is fatiguing in the manner already described. When the light falls directly upon the floor at the feet of the worker, and is reflected upwards into his eyes, the best plan, if the position cannot be changed, is to cover the floor with some dark stuff of a light absorbing quality.

For all who use their eyes continuously the conditions of illumination, although the most important, are not the only matters which require attention. The strength of the eyes may be impaired by overwork, as well as that of the body generally; and on this account it is desirable not to work to excess, and to take notice of any evidence of fatigue. Where circumstances compel continuous occupation, an endeavour should be made to vary it;

and, when this cannot be done, the eyes should be allowed
a few minutes' rest in the course of every hour, the
worker in the meanwhile changing his posture, perhaps
walking up and down, and at least directing the gaze to
large and comparatively distant objects. The time thus
spent, even though it may amount in the aggregate to an
hour in the day, cannot be considered as wasted, when
its value to the eyes is fairly taken into account. On
the same principle, that of avoiding overwork, the first
indications of presbyopia should receive attention, and
the necessary convex spectacles should be employed as
soon as ever the effort of accommodation begins to be
irksome. The manner of selecting these has been
described in a previous chapter ; but it may be permissible
to repeat here that they should be of sufficient strength
to afford complete relief, and that they should be so
adjusted, when there is no hypermetropia, that the eyes
may look over the tops of the glasses when raised for
the purpose of being directed to distant objects. In
commencing presbyopia, it is often said of spectacles
which are too weak that they "make the eyes ache."
The ordinary explanation of this is that they afford just
enough assistance to render it possible for the eyes to be
exerted to the point of over-fatigue. If they were somewhat stronger, so as to replace nearly the whole of the
accommodation effort instead of only a little of it, no
such fatigue would be experienced.

The *Purity and Temperature* of the air surrounding
them are of great importance to the eyes. This air may
be charged with dust, smoke, acrid vapours, or sharp

particles ; and may produce irritation or inflammation, either directly, or, when the eyes are tired and heated, by abstracting too much warmth from them. Against all such influences, however, the eyes are better protected than any other parts of the body, on account of their anatomical position in the midst of the brows, the lids, the lashes, and the prominent margins of the orbits. Notwithstanding these advantages, it is well known that they frequently sustain injury, even to the extent of being deprived of sight. It is therefore not superfluous to observe upon the means by which such disasters may to a great extent be prevented.

Persons whose occupations expose them much to dust should often wash the eyes with cold water. Clean river water, or, when it contains only a small quantity of mineral matter, clean spring water is the best for this purpose. Spring water which contains much lime should be boiled before being used. When the body is heated, care should be taken that the water is not too cold ; but washing with hot water relaxes the eyes, renders them more predisposed to inflammation, and as a rule, should only be practised in disease, or by the direction of a surgeon.

Small foreign bodies, such as the wing-cases of insects, particles of wood, straw, stone, glass, and the like, as well as particles of coke or ash from the fire of a locomotive, are often thrown forcibly against the conjunctiva by the wind. If they lodge upon the surface of the eye, or beneath the lids, the pain they cause speedily excites instinctive movements of the eyeball and lids, and a copious secretion of tears, by which, in many cases, the foreign body will be dislodged. When this does not happen, it will be found either on the cornea, or beneath the upper lid, a very short distance within its margin. In this position, the distress which it occasions may induce the person suffering to rub the closed lids with the fingers ; but in this way the foreign body may easily be driven more deeply into the tissue on which it rests, and its removal may thus be rendered more difficult. The lids should be kept open, either with or without the assistance of a finger, while the eye is rolled either up and down or from side to side ; and in this way the fragment may often be removed. If this plan should not succeed, washing with cold water should next be tried; and if this also should be unavailing, an attempt should be made to see the position of the foreign body by means of a hand-mirror, or it should be looked for by another person. If seen, the best implement for attempting its removal is the moistened corner of a soft handkerchief; and, if it still resists, the aid of a surgeon should be obtained, and until then, the lids should be kept as still as possible, and no treatment practised beyond some cooling application, such as a wet rag laid over the closed lids. When the foreign body is beneath the upper lid, it may often be removed by the sufferer, who should for this purpose take hold of the eyelashes of the upper lid and draw it away from the eye, and should then, whilst looking down, push up the lower lid, with a finger of the free hand, beneath the upper, so that the lashes of the lower lid may sweep the inner surface of

the upper one, when they will sometimes bring away
the foreign body as they descend.   If this plan fails, the
intruder may often be dislodged by passing the loop of
a glove-buttoner, or the central bent portion of a hair-
pin, along the inside of the upper lid ; but it must always
be borne in mind that after a substance has been removed
the sensation as if it were still there may nevertheless
remain for a time ; this being due either to a minute
wound upon the surface of the eye, or else to the fulness
of the blood-vessels, which quickly become distended
under the stimulus of any irritation, and which form
for a time actual prominent lines upon the surface.

The best remedy for this condition is generally the
application within the lids of a drop of castor-oil,
which not only diffuses itself over the eyeball as a
protective film, and thus shelters any abrasion from the
air ; but also, by its viscidity, fills up the hollows between
the temporarily prominent blood-vessels, and to some
extent restores the natural smoothness of the surface.
On the other hand, it not unfrequently happens that a
foreign body may remain undetected under the lid for
weeks or even months ; and that its presence there may
give rise to severe symptoms, which are all the while
erroneously attributed to some other cause.   Instances
of this kind have been seen by every ophthalmic
surgeon ; and they are only to be avoided by the very
careful inspection of every eye which becomes suddenly
irritated or inflamed.

Many artisans, such as smiths, ironworkers, millstone
cutters, stonemasons, and the like, are especially liable

to the impact of foreign bodies upon the eye; and, in all large shops for workers of such kinds, there is generally some one who has a reputation for skill and dexterity in removing impacted fragments. Nevertheless, as prevention is better than cure, it is wise, when the nature of the work allows, to protect the eyes by large spectacle frames, the rings of which are filled up by a suitable material. For stone-breaking or other work which does not require minute vision, this material may be wire gauze of sufficient fineness; and, where gauze would interfere with sight, the best substitute may be found in plates of thin talc. Such talc spectacles are made in Germany at very small cost; they are quite transparent, and are strong enough to resist any blow to which they are likely to be subjected. They are also large enough to shield the eyes not only from direct impacts, but also from bodies flying towards them in an oblique direction.

When a foreign body does not rest upon the surface of the eye, but has actually penetrated its tunics, the case becomes one of a highly serious character, not only for the eye first injured, but also for its fellow. Injuries of a certain class are liable to excite, in the uninjured eye, a form of inflammation which is called sympathetic ophthalmia, and which is of a most intractable and destructive character. In such circumstances, it is often necessary entirely to remove the injured eye in order to preserve the sound one; and to do this before sympathetic ophthalmia has actually made its appearance. Such cases, in all their stages, require skilled surgical treatment ; and they are only mentioned here in order that their extreme gravity may be insisted upon. They are often caused by pellets of shot, fragments of guncaps, the bursting of bottles containing aërated liquids, by thrown stones, and especially by a dangerous toy called a " cat," with which street children are accustomed to play at certain seasons of the year. I have seen many poor people blinded by these "cats," and some of them have lost the second eye also from sympathetic ophthalmia, because they could not be convinced of the risk which they incurred, or induced to have the injured eye removed in time.

Another class of accident is that in which the eyes are injured by caustic or heated fluids, as by strong acids, boiling water, melted metal, or soft mortar or other mixtures containing lime. In such circumstances, while medical aid is being procured, any of the noxious material which can be seen within the lids should as far as possible be removed or gently washed away ; and a little oil, a soft poultice, or other soothing covering, should be used as a temporary protection. Such cases range from the most trifling injuries to those which occasion complete destruction of the eyeball.

A case has been recorded by Beer which is worth quoting for the sake of the warning which it affords. He was called to a man whose eyes had always been good, but who was blinded in an instant by a foolish trick. The man was in company with friends, when some one came into the room behind him, and placed both hands over his eyes, at the same time bidding him guess who was doing so. The man either could not or would not guess; and, at all events, he struggled to free himself. The newcomer tightened his grasp, the eyes were open, or were forced open in the struggle, and the person thus assailed was irretrievably blinded by the pressure of the assailant's fingers.

Since the eyes are integral parts of the organism as a whole, it follows that their comfort is inseparably bound up with its general welfare, and that they are liable to share in all morbid conditions either of body or mind. Of their participation in actual disease it is scarcely necessary to speak ; otherwise than by recalling how much the capability of using them, in so far as it is dependent upon the powers of accommodation and convergence, may be diminished by states of general muscular weakness, in whatever way these have been brought about ; while there are certain maladies, especially of the brain, the spine, and the kidneys, which affect the eyes in a more direct and special manner. Apart from disease or bodily weakness, the eyes are also affected by moral states, and by the emotions, more particularly by those of a depressing character. It is a common saying that the eye is a mirror of the mind ; and nothing is more remarkable than the rapidity with which mental states modify the quality and the amount of the secretion of the corneal surface, causing this surface to look either bright, or dry and dull, according to their characters. Anger, grief, and the depressing emotions generally, especially states of chronic worry, lower the whole tone of the circulation,

Q

and impair the nutrition of all parts of the body. The eyes not only share in the general effect, but they are almost the first organs in which it becomes manifest. They become sunken, dull, and incapable of their customary amount of application. Continued or habitual weeping also acts upon the eyes very injuriously, not only in the same way as depressing emotions generally, but also by producing redness and irritation of the lids and chronic congestion of the surfaces. The distended blood-vessels do not immediately return to their natural size ; and the internal structures of the eye share, more or less, in the over-congestion of its superficial parts.

The smoking of tobacco is hurtful in all cases in which the eyelids are sore or irritable, as would be the exposure to smoke of any other kind. Of course, this effect is more likely to be produced in a railway carriage, or in a room with other smokers, than by solitary smoking, or by smoking in the open air. To smoke for hours together whilst reading or writing, especially by artificial light, can hardly fail to be prejudicial as a matter of surface irritation ; and the congestion thus produced may extend to the interior of the eyes. Beer expressed the opinion, more than sixty years ago, that smoking in youth, on account of the frequent loss of appetite which it occasioned, was often a cause of permanent weakness of sight ; and the same notion was again promulgated, in a somewhat different form, about eighteen years ago, when an attempt was made to trace atrophy or wasting of the optic nerves to the use of tobacco. Concerning this doctrine, I will say no more than that it seems to me not only to be not proven, but even to be opposed to facts about which no doubt can be entertained. The consumption of tobacco has increased enormously in this country within the last twenty years, particularly among boys; and I at least have not observed such a corresponding increase in the prevalence of nerve atrophy as there should have been, if the belief of its dependence upon tobacco were well-founded. In saying this, I speak quite without being influenced by any partiality for smoking, seeing that I do not practise it, and that to-bacco, in all its forms, is rather distasteful to me than otherwise.

The use of snuff, now practically almost obsolete, was at one time supposed to be beneficial to the eyes. When the eyes are healthy, it is a matter of absolute indifference whether snuff is used or not ; and I am not aware of any facts which show that remedies can be usefully applied to diseased eyes through the medium of the nostrils.

The consumption of alcoholic drinks, whether wine, beer, or spirits, has been lauded and condemned, with reference to the eyes, without much, if any, evidence either way, by the more heated advocates or equally heated opponents of total abstinence. Some writers, especially in Germany, have made as resolute an effort to connect atrophy of the optic nerves with the abuse of alcohol as some of our own countrymen have made to connect it with the abuse of tobacco. Among the not very large number of persons who suffer from optic nerve atrophy, of a kind traceable neither to prior inflammation, nor to disease of the brain or spinal cord, there can be little doubt that some have been intemperate. Optic nerve atrophy, occurring in an intemperate person, like any other malady which may afflict or punish him, is probably at least so far connected with his vice that he would have been less likely to suffer from it if he had been a sober man. When I have admitted this I have admitted all, as far as I am aware, which can be said to be sustained by any kind of evidence. To go farther, not only with reference to nerve atrophy, but also with reference to other forms of disease, more especially, perhaps, to insanity, is not only to lose sight of scientific precision and exactness, but also, in many cases, to incur a risk of libelling the unfortunate. People who are unaccustomed to reason hear that a large proportion of the inmates of an asylum have been intemperate before they were placed under restraint ; and they cry out : " See how drink is a cause of insanity ! " They forget that the other view is at least equally tenable ; and that the case may be as fairly stated by saying, " See how insanity, in its earlier stages, is a cause of drunkenness ! " The man who would preserve the full integrity of his functions to a ripe old age must avoid excesses of every description, and must endeavour to employ the higher faculties of his mind somewhat more energetically than is now always customary. A time comes to every one when the physical powers begin to decay ; and then, unless the brain has been kept active and recipient by exercise, there is nothing left to live, and the man perishes. We say that he died of gout, or of over-eating, or of heart-disease, or of kidney-disease, or of the failure of the particular organ which was the first to exhibit symptoms of the approaching end.   In reality, he has died of stupidity, artificially produced by neglect of the talents with which he was endowed. That which is true of the organism as a whole is true also of its parts; and the eyes, among others, are best treated by an amount of systematic use which preserves the tone of their muscles and the regularity of their blood supply.   The acuteness of sight, moreover, is in a great degree dependent upon the mental attention habitually paid to visual impressions; and I have often observed this acuteness to be below the natural average in agricultural labourers who, if able in some sense to read, were not in the habit of reading, and who were not accustomed to look carefully at any small objects.   I have even had reason to think that the wives of such men were indebted to their household needlework for the maintenance of a higher standard of vision than that of their husbands; and I have no doubt that idleness of the eyes, if I may use such an expression, is in every way hurtful to them, and that proper and varied employment is eminently conducive to their preservation in beauty and efficiency.

CHAPTER XII.

IT happens, with quite a curious frequency, that men and women are placed in positions, or are desirous to undertake occupations, for which the condition of their sight renders them exceedingly ill-suited ; and it also happens that persons, whose vocations were originally such as they could fulfil without injury, may be temporarily or even permanently disabled from following them on account of defective sight. It therefore becomes important to know what can be done to save eyes which are overtaxed, either by calling some other sense to their assistance, or by diminishing the stress which would ordinarily be thrown upon them ; and it is not a little remarkable that some of the greatest monuments of human industry have been at the same time victories over disabilities which would have prevented most persons from even attempting to exert their faculties in such a manner. Prescott, the historian, for instance, was so nearly blind that much reading was impossible to him, and even writing was a task of great

difficulty ; and he was probably the pioneer of all subsequent inventors of appliances by which writing can be
accomplished with little or no assistance from vision.

Some two or three years ago, Dr. Thursfield, the
Medical Officer of Health for Shrewsbury, was suffering
from an affection of the eyes on account of which he
was advised to abstain from all employment of them
over near objects ; or, in other words, from all reading
or writing. His visits of inspection into sanitary matters
were still permissible; but the employment of a reader
and of an amanuensis seemed to be essential to enable
him to retain his office. The reader was an admitted
necessity, from which there was no escape ; but the
amanuensis wearied him. He made inquiries for such
a machine as Prescott had employed ; but found nothing
which was entirely satisfactory ; and he thereupon set
to work to construct one for himself. In the course of
a short time, he succeeded in producing a frame which
enabled him to write without using his eyes ; and this
frame, after being on two or three occasions improved in
points of detail, at last assumed the appearance which
is shown in Fig. 49. It is not only calculated to do for
others what it did for its inventor, but also to allow
blind people, who have become blind after they have
been accustomed to write, to continue the practice
without difficulty. In the treatment of certain forms of
eye-disease, it is an adjuvant of great value; since it
lightens the tedium which would otherwise attend upon
enforced idleness ; and it has the additional merit of so
steadying the hand that it will enable seeing people to

write straight and legibly in a railway carriage, however great the speed of the train, or however disturbing its oscillations. It can be used either with a pencil and prepared "metallic" paper, on which the writing is indelible; or with a style and carbon paper; or with a

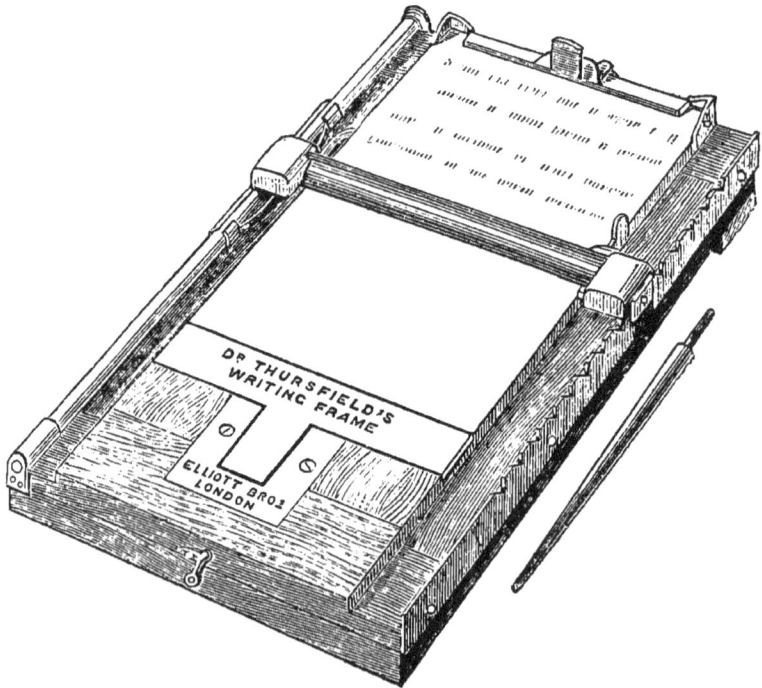

FIG. 49.

" stylographic pen " and common paper; and such a frame is now my constant companion on a railway journey of any length. It is as easy of use, after a little practice, by night as by day; in a tunnel as on an embankment. It consists essentially of a board to write upon, with a transverse bar to guide the fingers; and this bar slips down, when the style or pencil reaches its right hand extremity, by one division of a sort of rack, so as to preserve always the same space between the lines. The successive improvements which Dr. Thursfield has made have been in the mode of attachment and of sliding of the bar, in greater facilities for inserting and withdrawing paper, and in converting the bed of the instrument into a shallow box, which contains a sufficient supply of paper and of styles or pencils for a considerable amount of writing. By the aid of this machine, after a few attempts with the eyes shut, it is possible to write a perfectly regular hand without any aid from sight, and therefore without any strain upon the eyes ; and a very fair rate of speed is soon attained. Apart from its other advantages, it offers to those who are wakeful at night facilities for writing in bed without the use of a lamp ; and thus furnishes them with a resource which may often serve instead of reading, and which is wholly unobjectionable as far as the eyes are concerned. Thursfield's writing frame is made by Messrs. Elliot, the Philosophical Instrument Makers in the Strand, opposite the Charing Cross Railway Station.

Another contrivance, and one of more general value and applicability, is the American type-writer, shown in one of its best forms in Fig. 50, manufactured by Remington and Sons, at Ilion, in the State of New York, and sold in London at 54, Queen Victoria Street. In my own hands at home, writing a very great deal, both for publication and in correspondence, this machine has for the last four years superseded the pen entirely, so

that I now use the latter for no other purposes than to
correct proofs by interlineation, to sign my name, and to
address letters. The type-writer is worked by vertical

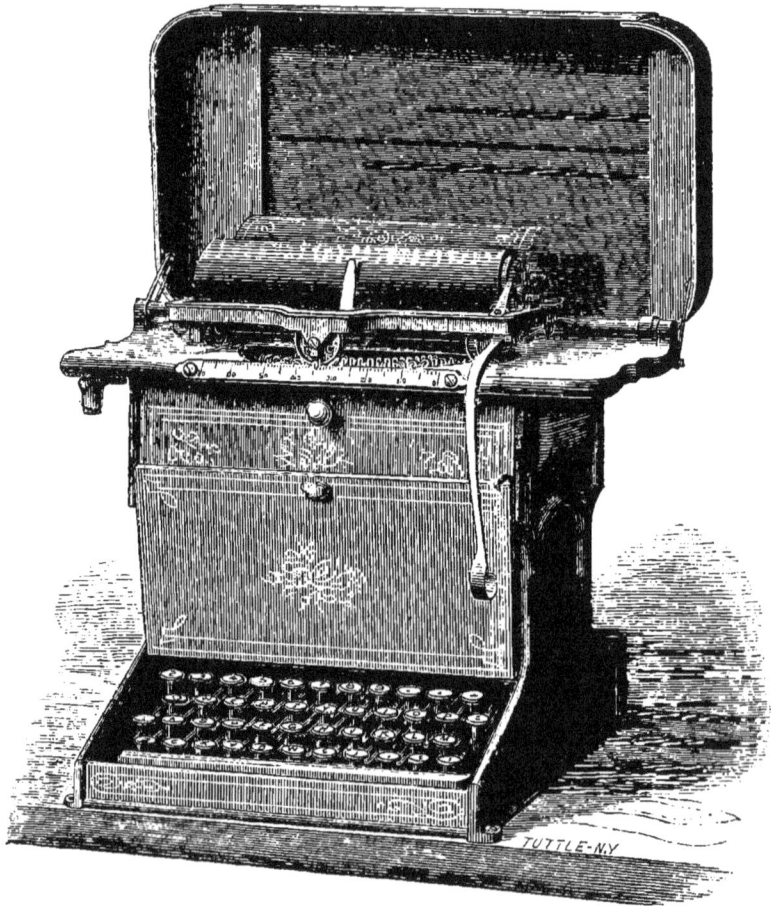

FIG. 50.

keys, like those of a cornet-a-piston, having finger
surfaces five-eighths of an inch in diameter, on which
the letters or figures are enamelled in characters five-

sixteenths of an inch high. The key-board contains forty-four keys, in four rows of eleven ; and the arrangement is not alphabetical, but is like that of the compositor's box, in which the letters are placed nearer to the centre in the order of the frequency of their occurrence in words.

Even in learning the use of the instrument, when the letters have to be looked for one by one until the fingers become sufficiently familiar with their several positions to touch them instinctively, the characters are so large that there is no appreciable strain upon vision ; and, when once dexterity is attained, the eyes can scarcely be said to be used at all. My own eyes have never occasioned me any discomfort; and my own use of the type-writer rests upon quite different grounds ; but yet my experience of it enables me to recommend it, very strongly, to all persons who, having to write much, are made conscious by the exercise that they have eyes. For the short-sighted it is especially valuable ; because there can never be any inducement to stoop over it, so that a great snare to them in writing is altogether set aside. Next only in advantage to the facilities it affords for writing are those which it affords for reading what has been written ; for this is printed in perfectly spaced lines, at regular and if desired at rather wide intervals, in block capitals of perfect clearness. The author who wishes to glance back over his MS. is almost as much helped as he who only wishes to produce it ; and the labour of seeing, in every stage of the process, is either abolished or reduced to a minimum. So much is this the case that

I am informed, although I cannot speak on this point from personal knowledge, that its use is readily acquired by the blind.

The other advantages of the machine, those for which chiefly I personally value and employ it, are worthy of a moment's consideration. Its work in my hands is about twice as rapid as that of a pen; and became so after a few weeks of practice. A new machine is frequently a little stiff, as with machines of every kind; but, when once the type-writer has been worked into running order, the keys move as easily as those of a piano, and the effort of tapping them, being accomplished by the finger tips, and being divided between two or three fingers of each hand, is unfelt, however long it may be continued. I have worked the machine for eight consecutive hours without more than ten minutes interruption, and at the end of that time my hands were not conscious of the least fatigue. Every writer is aware that the same thing cannot be said with regard to the pen : for that the effort of holding it, which is performed almost exclusively by the muscles of the ball of the thumb, opposed by those of the fore and ring fingers, soon becomes very fatiguing; and, in professed copyists, not seldom produces an affection known as scrivener's cramp, which sometimes passes on to complete paralysis of the muscles which are concerned. Before I possessed a type-writer, my own muscles were often so tired by writing that I had seriously thought of abandoning the practice, lest the strain thrown upon them might in time impair the qualities of the hand as

an instrument for surgical purposes; but since I have
had a type-writer, these feelings of fatigue are unknown,
not only on account of the division of labour between
several fingers, but also because the work is not done by
the small muscles of the fingers at all, but by the large
ones of the fore-arm, which are capable of much greater
and more sustained exertion.

Of less importance, but still not to be left out of
account, are the beauty and clearness of the MS. which
the instrument produces. All authors are aware that
it is impossible to put the finishing touches to written
composition until it appears in the form of a proof;
because the comparative illegibility of even the best
handwriting prevents that looking forward which is
essential to complete correction. Hence it follows
that the cost of corrections forms an important item
in the cost of publication. With a type-writer, this
difficulty no longer exists. The result which it produces
is clearer and more legible than any ordinary proof;
and every correction can be made, by interlineation
either with the machine itself, or with a pen, before
the MS. is sent to the printer. The fair copy of the
present treatise, for example, was produced by my
own hands in an aggregate of about fifty hours; and
almost the only faults which required correction in the
proof were the few accidents from wrong types which
had been overlooked by the reader. The original
setting-up was in the pages as they now appear,
instead of in slips as usual; and the total cost of
a type-writer would be more than repaid, to many

authors, by the saving which it would effect in the production of even a single volume.

Like other machines, the type-writer did not spring full-grown from the brain of its inventor, but has only gradually been brought to its present degree of excellence. Since I first obtained one, several improvements of detail have been introduced ; and one of the latest forms is constructed to print either capitals or small letters at the will of the operator, instead of capitals only. Another form is fitted with a plate-glass cover, which incloses the whole of the working parts, and intercepts all sound. Without this cover, some tapping noise, comparable to that of some of the forms of sewing-machine, is an inseparable accompaniment of the instrument. The glass case is said to have another advantage, in that it prevents the inking ribbon, with which the machine is supplied, from being readily influenced by changes in the hygrometric condition of the atmosphere. Without this protection, the ribbon is soon affected either by dryness or moisture; and hence, for continuous work in changeable weather, it is well to keep two ribbons, one somewhat more saturated than the other.

Upon all the above grounds, and after four years of daily personal use, I most strongly recommend the type-writer to all persons who write for considerable periods of time, more especially to the short-sighted, or to those who have any kind of trouble about their eyes. Even apart from such trouble, I think no systematic writer should be without it, unless he be one of

those curiously constituted creatures to whom the caprice
of nature has denied the gift of mechanical aptitude,
even in its most rudimentary form. Those who possess
two left hands, and whose fingers are all thumbs, had
better, perhaps, leave the type-writer alone ; but for all
others it is an inexpressible relief either for weary hands
or for weary eyes ; and its prime cost will be speedily
repaid by the resulting economy both of money and of
time. The only kind of writing to which it is inapplic-
able is for making entries in books ; and even this difficulty has to a great extent been overcome in America
by the practice of using separate sheets, which are preserved in a portfolio and then bound in a volume. The
records of the ophthalmic department of St. George's
Hospital are now being taken by a type-writer and
preserved in this manner.

In using the type-writer, it is quite possible to keep
the shoulders square and to sit erect ; but nearly all
persons who write with a pen at a low table or desk
contract a habit of stooping, which is in many ways
prejudicial to them. It not only contracts the chest, so
as to interfere with the freedom of respiration and thus
with the due aëration of the blood ; but it also tends to
produce congestion of the head generally and of the
eyes in particular. Those who are constrained to write
with a pen should keep their work at a somewhat high
level, and should endeavour, as far as may be possible,
occasionally to exchange the sitting for the erect posture.
The best way of doing this is to have a standing desk,
on which the books or papers may be placed; but it

would not be desirable to write standing for many consecutive hours, on account of the swelling of the legs
and the varicosity of their veins which would almost
certainly be produced. Many workers, who desire to use
a standing desk occasionally, may perhaps be helped by
one of American invention, and of very ingenious construction, which has been introduced into this country
by Messrs. Thomson and Sterne, of Glasgow and of
9, Victoria Chambers, Westminster. This desk admits,
when not in actual use, of being folded and put away
into a space almost inconceivably small; and, when
it is set up, it affords ample space for all ordinary
requirements.

As regards writing, as well as regards all other occupation of the eyes upon near and small objects, the
rules already laid down, with regard to the nature and
the position of the light whether solar or artificial,
should be carefully observed ; but in writing we may
supplement these rules by the advice to make the letters
as conspicuous as possible. Short-sighted people are
especially prone to fall into a habit of small and nig-
gling writing, against which they should strive persist-
ently, always remembering that such writing is both a
consequence and a cause of that approximation of the
eyes to their work, which tends surely to the increase
of their defect, and which they should be always watch-
ful to avoid. In order to render writing conspicuous, it
is generally desirable to use paper which is either white
or cream-coloured (although persons with very sensitive
eyes may prefer a faint tint of blue), and to write upon

this paper with ink which runs black from the pen, and which thus presents a bold and conspicuous contrast to the ground on which it stands. By using such ink, and by tracing with it bold and well-formed characters, the writer will not only preserve his own vision, but will also earn the gratitude of those who are called upon to read his writing. Cobbett long ago laid down the simple but much neglected principle that the object of publishing a book is that people may read and understand it ; and he showed how often this natural object was lost sight of, or was even wholly defeated and rendered unattainable, by neglect of the most simple principles of grammar. In like manner, letters and MSS. are presumed to be written in order that people may read and understand them ; but there are many scribes who fail to fulfil the primary conditions of the requirement. I do not recognise the right of any one to add to the burdens of life, and to the inevitable stress which falls upon the human eyesight, by writing clumsily and illegibly ; and not the smallest of the boons which the type-writer has conferred upon humanity is in furnishing us with manuscript in clear and legible types, instead of with the hieroglyphics which were wont to proceed from the same sources prior to the introduction of the machine.

As regards the production of written or printed characters, we have, therefore, many means by which the labour of the eyes may be either greatly lightened or even altogether set aside ; but as regards deciphering these characters we have no similar resources. This

R

difference alone is sufficient to render it desirable that
students and others, whose working hours are divided
between reading and writing, and extend into the night,
should endeavour, as far as circumstances will permit,
to read in the daytime and to defer their writing until
night. Besides this, the only things which can be done
to facilitate reading are to choose a position in which
the light falls from the right direction and at the right
angle upon the page, and to place the book at such a
height and in such a manner as to avoid all constrained
or bent postures of the neck, by which the free return of
blood from the head may be impeded. For this purpose some of the ingenious contrivances of Mr. Carter,
of New Cavendish Street, will be found useful, since they
support a book and a light in almost any desired position before the eyes. Those whose eyes are in any way
troublesome should also make it an invariable rule to
reject very small or badly-cut type, and to read only in
the largest type in which the books which they wish
to study can be procured. Students and authors will
every now and then be engaged in work, such as com-
pilation or translation, which requires reading and
writing too ; and in these circumstances they should
be careful to place the authority which is being referred
to, or the book which is being translated, as nearly as
possible at the same distance from the eyes as the paper
on which they write, and as nearly as possible under
the same degree of illumination. When engaged myself
in work of this description, I have found it useful to
have two or three small reading-desks, each capable of

supporting one volume, and each standing upon a short central leg with a weighted foot. These desks were ranged round me in a semicircle on the table, equidistant from my eyes and from the lamp, and could all be referred to with equal facility. Details of this kind have an appearance of being trivial; but it will be found in practice that upon such trivialities comfort very largely depends.

CHAPTER XIII.

IN addition to the various references which are scattered through the foregoing pages, there are certain points connected with spectacles which it may be convenient to bring together into a single chapter. The general term includes two principal divisions or classes : spectacles properly so called, and " folders ; " a " folder " being also known by its French name of "pince-nez."

In spectacles proper, the lenses are contained in a frame which is intended to secure them to the sides of the head, and which is shown, in its simplest form, in Fig. 51. The frame consists of three distinct parts ; the eyes, or rings ; the bridge ; and the sides ; and it is usually made of gold, silver, or steel, the latter being sometimes plated, nickelled, or gilt. The sides may be single, as in the figure, or double with a turnpin joint, as in Fig. 52, or "curled" as in Fig. 53. The object of the curled or the double side is to pass round behind the ear, and to hold the lenses in some fixed position before the eyes.

Folders are of two chief varieties, those like Fig. 54, in which the two parts are connected only by a rivet, which

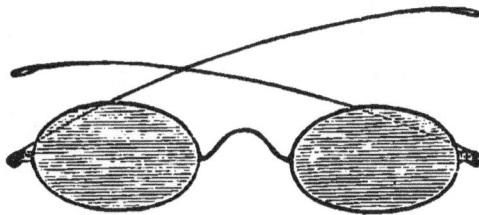

FIG. 51.

should be sufficiently tight to give them a grip upon the nose ; and those like Fig. 55, in which the two parts are

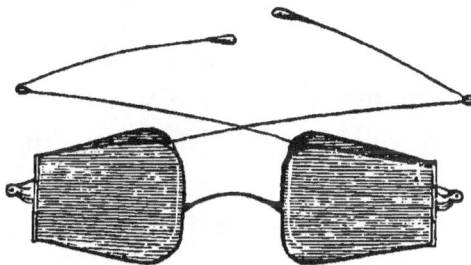

FIG. 52.

connected by a spring.   Those like Fig. 55 are known by makers as the English pattern, in opposition to Fig. 56,

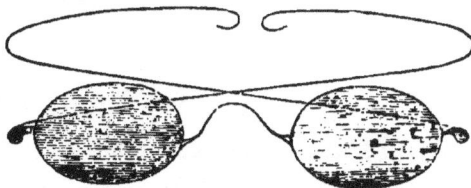

FIG. 53.

in which there are additional pieces to keep the lenses farther away from the sides of the nose, and which is

known as the Japanese pattern. Folders are commonly
made of steel, aluminium, gold, or tortoiseshell ; and the
English pattern may be frameless, the handle and the
connecting spring being riveted into holes drilled through
the lenses. There are various unimportant modifications
of these patterns ; and there are also mounts which are

 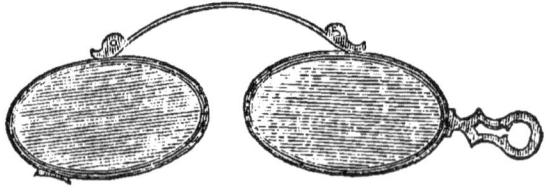

FIG. 54.   FIG. 55.

something intermediate between a folder and a spectacle
frame, which are made with a distinct bridge between
the glasses, and which shut up into some kind of handle.
It can scarcely be said that these latter forms have any
advantage, beyond the larger profit which costliness of

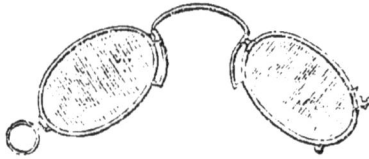

FIG. 56.

material and of workmanship may afford to the dealers
by whom they are sold. The essential point in all such
appliances is in the quality and fitness of the lenses, and
a mount which engages one hand, and leaves but one
free for any other kind of occupation, can scarcely be
anything but a disadvantage.

In order to consider the respective merits of spectacles
and of folders, it is necessary to remember what it is
that they are required to do.   The object, in every case,
is to hold lenses before the eyes, at a distance and in a
position to render them aids to vision ; and the distance
and the position will somewhat vary according to the
kind of employment which is being pursued.   There is,
however, one universal requirement ; namely, that the
eyes should look through the centres of their respective
lenses in every instance in which decentration is not
designedly and in a definite degree prescribed.   It has
long been known to opticians, as a matter of empirical
observation, that the discomfort arising to the presbyopic, from the premature use of glasses of too high a
power, is much increased if these are mounted in frames
a little too wide for the patient, so that the eyes look
through the inner sides of the lenses instead of through
their centres ; and also that the discomfort is diminished,
or is less likely to arise, if the frames are somewhat
narrow.   When the relation between accommodation
and convergence became known, the action of misfitting
frames was rendered easily intelligible.   A convex lens,
within its curved surfaces, has already been described as
made up of an infinite number of prisms with their bases
meeting at the centre ; and a concave lens is in like
manner made up of an infinite number of prisms with
their bases outwards or at the margin.   Hence, as shown
in Fig. 57, a person who looks through the inner sides
of convex lenses, as he is compelled to do whose frame
is too wide for the distance between his eyes, is looking

not only through convex lenses, but also through prisms
with their bases outwards; and he who looks through
the outer sides of the lenses, as must happen when the

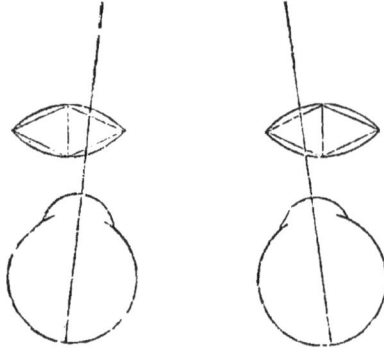

FIG. 57.

frame is too narrow, and as shown in Fig. 58, looks
through prisms with their bases inwards.   With concave
lenses, of course, these conditions would be reversed.

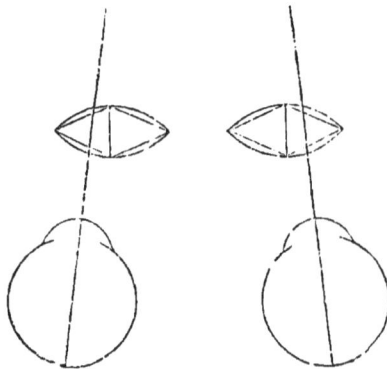

FIG. 58.

The former arrangement calls upon the convergence
muscles for increased effort at a time when the accommodation muscles are relaxed ; while the latter rests the

convergence together with the accommodation. The former must at all times be a source of discomfort, the latter is only comfortable when the strength of the lens and the degree of its decentration are in perfect harmony. A simple illustration of the effect of a misfitting frame may be obtained by taking a lens, closing one eye, and looking through the lens, held four or five inches in front of the other, at any moderately distant object. If the lens be now slightly moved in and out across the eye, the object looked at will appear to move also ; and the manifest explanation of this appearance is that we only see an object in its actual position when looking through the centre of the lens, and that we see the object displaced, or in a false position, when looking through any of the lateral parts of the lens. If, then, we put before the two eyes glasses each of which is out of centre, we shall give each eye an image which is more or less displaced ; and the two eyes, in order to fuse these two displaced images, or to combine them into that of a single object, must squint in a degree corresponding with that of the displacement. The effect of this necessity is to disturb the natural harmony between the movements of the two eyes, to force each of them into a strained and unnatural position, and to fatigue and distress the muscles by which this unnatural position is maintained. I have known several instances in which discomfort in wearing spectacles, so considerable as to be prohibitory of their use, has been produced solely by misfitting frames, and has been removed as soon as lenses of the same power were put into frames adapted to the requirements of the wearer. It must be remembered that it is the centre of the lens, the thickest part of a convex and the thinnest part of a concave, which should be in front of the pupil of the eye, and not necessarily the centre of the ring of metal in which the lens is mounted. In cheap spectacles, there is often an accidental decentration of the lens which renders its centre by no means coincident with that of its setting; so that I have seen instances in which an apparently misfitting frame was comfortable because its lenses were accidentally out of centre, and other instances in which the reverse of these conditions obtained.

It was suggested by Dr. Giraud-Teulon, of Paris, many years ago, that the systematic decentration of spectacle lenses might be made to serve important uses; the prismatic element which would thus be introduced being made to regulate the convergence, while the convex or concave element regulated the accommodation. The idea was afterwards worked out very completely by my friend Dr. Scheffler, of Brunswick; at whose request, in 1869, I translated into English his treatise upon the subject, with large manuscript additions from his hand, and thus, as mentioned at page 145, introduced the use of prismatic or decentred glasses to the profession in this country. Regarded merely from the standpoint of physiological optics, they seemed likely to be valuable, but when tested by the actual demands of life they were extremely disappointing; so much so, indeed, that I have long ago abandoned their employment save in a few exceptional cases. I am only acquainted with one

ophthalmic surgeon in London who systematically pre-
cribes them ; and I have frequent evidence in my con-
sulting room that they fail in the cases of his patients as
completely as they failed with my own.   Dr. Scheffler
himself, unfortunately, is not a practitioner ; and his
researches into the question were matters of mathematical analysis and calculation, from which, as experience soon demonstrated, certain important data had
been omitted.

Returning from this digression to the requirement
that the centres of spectacle lenses should be opposite
to the pupils of the eyes, or, more properly, that each
centre should be in a line connecting the centre of the
pupil with the object looked at, we have to consider,
first, that the pupils themselves will be brought nearer
together by convergence in proportion to the nearness
of the object, and, secondly, that convex glasses which
are used only for near objects may be placed somewhat
below the eyes, and that their power is increased if they
are a little distance away.   The nose, on which a folder
takes its grip, consists of bone at the upper part and of
elastic cartilage lower down ; so that a pince-nez for
reading may be placed upon the latter part, where the
compression which causes it to retain its hold serves
also to press the nostrils together, and thus not only
checks inspiration, and renders it necessary to open the
mouth to breathe (in itself a highly objectionable practice), but also alters the tone of the voice.   A person
attempting to read aloud, when the nostrils are compressed by a tight pince-nez, produces very ludicrous

sounds ; and, if the pince-nez is not tight, it is apt to
be dislodged by the muscular movements of the face,
and to fall off in the middle of the performance, the
owner having to find and to replace it before he can
resume the interrupted sentence.   In reading during a
solitary meal, again, a pince-nez is apt to be loosened
by the movements of mastication, and to fall off into
the plate.   As regards the wants of the presbyopic,
therefore, I recommend spectacles in all cases. .They
may be of the simplest pattern, with single sides, which
should be long enough to allow the glasses to come
nearly half-way down the nose, and yet to hold them
there securely.   In this position, when the eyes are
raised to look at any distant object, they will look
quite over the glasses without effort ; and, when cast
down to a book or other small object of vision, held in
the hand or lying upon a table, they will look through
the glasses by virtue of the mere change of direction.
Moreover, when spectacles are required for reading only,
the rings containing the lenses should be set at an
inclination to the sides, so that the lenses will lean
forward at their upper portions, and will present surfaces at right-angles to the rays of light which proceed
from the surface looked at.   The uses of a pince-nez
in such cases are merely supplementary.   It may be
conveniently worn attached to a chain or cord, round
the neck or fastened to some part of the dress, so that it
can never be lost or left behind ; and the owner is then
secured against forgetting his spectacles, an accident to
which most persons are liable on first beginning to use

them. When thus worn a pince-nez is more ready of
access than the spectacles themselves, which should be
in a case and in the pocket when they are not upon the
nose ; and it may therefore be a convenient resource for
reading any short document, or for inspecting any small
article. It has another important use by affording a
useful occasional addition to the spectacles, and may be
put up in front of them in a bad light, or to look at
something unusually small or illegible, such for instance
as the figures in the columns of "Bradshaw." Regarded
as an accessory, there are many circumstances in which
it is highly useful, but it should never be made the chief
reliance of its owner. A pince-nez may be most conveniently worn attached to a very slender chain about
fifteen inches long, furnished at the other end with a
hook which can be passed through an eyelet hole in the
right shoulder of the waistcoat. In this position, when
it is suffered to drop, it falls out of harm's way against
the side of the wearer ; whereas, if it is worn by a cord
or chain round the neck, so that it hangs in the middle
line of the body, it is apt to chink and jingle against the
watch-chain, and is much exposed to be accidentally
broken against the edge of a table. The chain should
be very slender, as otherwise its weight tends to drag
the glasses off the nose.

All persons who require glasses for distant vision,
whether they are myopic or hypermetropic, require
these glasses to be held near to the eyes and at a high
level, so that the eyes look through, and not above
them, when directed straightforwards at distant objects.

There are some people in whom the bones of the
upper part of the nose are so shaped that a pince-nez
will obtain a good hold of them, and will keep the
lenses in the required position, but such people are in a
minority ; and, generally speaking, spectacles are the
best things for permanent wear. The various shapes
which may be given to the bridge of a spectacle frame
render it easy to have the glasses held up as high as
may be required ; and the invariable length of the
bridge keeps these glasses always at the right distance
apart, and always, therefore, correctly centred, supposing that they were so at the beginning. The curled or
turn-pin sides, moreover, afford an additional source of
stability ; and enable the wearer to run, ride, dance, or
perform any other movements, without the glasses
becoming displaced. Where a pince-nez can be worn
permanently, that is, where it will fit the nose, the
Japanese pattern will generally be the best, on account
of its serving to keep the lenses apart ; and a variety of
contrivances have been introduced for the sake of diminishing its pressure without detracting from the security
of its hold. Some of these contrivances are very ingenious and moderately successful ; but I confess that
I greatly prefer spectacles to the pince-nez for every
kind of permanent use, and I think the very general
popularity of the latter is due to the fact that many
people look upon declared spectacles as a sort of
badge of infirmity, while they consider a pince-nez to be
a mere eye-glass, an airy nothing, which it will be supposed they might lay aside if ever they were inclined to

do so. The delusion is a harmless one, as long as it does not bring the eyes into any serious trouble.

The general fact is, as will be seen from the foregoing, that spectacles are better than folders because they can be more perfectly fitted to the wearer, and can fulfil more certainly the individual requirements of his case. Folders, as a rule, are to be regarded only as more or less ill-fitting spectacles. The discomforts produced by ill-fitting spectacles do not become prominent unless the spectacles are used continuously; and hence folders have certain advantages for use during brief or occasional periods of time. For all kinds of continuous employment, spectacles are greatly to be preferred. The wearer of spectacles, however, should take care to secure all the benefits which they are calculated to afford, and for this purpose he should pay as much heed to the fitting of the frames as to the suitability of the lenses. Nothing is more surprising in practice than to see the way in which people will wear the spectacle frames of others larger or smaller than themselves; the way, for example, in which a wife will wear her husband's glasses, and will wonder at the aching and uneasiness which they occasion. If the general principle of the importance of a well-fitting frame were properly understood, the complaints of discomfort arising from spectacles would be much less frequent than at present. The fitting of the frame should, of course, be as much the business of the optician who sells it as the fitting of a coat is the business of a tailor; but this important element is often as much neglected by the seller as by the purchaser. On the good old maxim of *caveat emptor* it should at all events be the business of the latter; and every buyer of spectacles, before taking them, should put them on and should look at himself or herself in a mirror. If the spectacles are for distant objects, the eyes, when looking horizontally into space, should have their pupils exactly in the middle of the rings ; and, if for reading only, the pupils, in the same position, should be a little nearer the nose than the centre of each ring. The purchaser should also require from the seller a warranty that the lenses are correctly centred ; that is to say, that the centre of each lens, its thickest or thinnest part, is in the centre of the ring in which it is set.

There are many people who require glasses of different powers for different purposes. A hypermetropic person, for example, will require, when he reaches the age of commencing presbyopia, a stronger glass for reading than for distance ; and a myopic person will frequently require a weaker concave for reading than for distance. In all these cases, it is often possible to avoid the inconvenience and loss of time incidental to frequent changes by combining the two powers in one and the same frame. The suggestion to do this was first made by Benjamin Franklin ; and hence the spectacles in which it is done are commonly known as Franklin spectacles. Glasses of the power required for distance, and glasses of the power required for near objects, are selected, and are cut in two horizontally, two dissimilar halves being combined in each ring of the spectacle frame. The half glasses for distance are put in the upper parts of the rings, the half glasses for near work in the lower parts. The wearer therefore looks through the former when the eyes are directed forwards, through the latter when they are directed downwards. The fine horizontal line of union is quite imperceptible to the wearer, who always looks either above or below it; but, in order to make such glasses quite comfortable, the division of the lenses must not be carried through their optical centres, but sufficiently above or below them to bring them in front of the pupil in each position. If this were not done, some apparent displacement of objects would be occasioned. A similar effect has been produced, originally by French opticians, by grinding two different powers upon one and the same lens, so as to avoid the appearance of a crack or line along the horizontal diameter of each of the glasses. These *verres à double foyer*, however, are decidedly inferior to the compound lenses made by Franklin's method, because, in the former, the line of demarcation between the two powers is apt to be irregular in its outline. For all double glasses, of course, and also for the cylindrical glasses required for astigmatism, and generally for all in which the maintenance of the lenses in some given position is required, the use of spectacles as opposed to folders is essential to the attainment of the best results. Frames with curled sides are generally to be recommended in such cases, but it is necessary that they should fit accurately not only as regards the shape of the bridge and the distances between the centres, but also as regards

S

the exact length of the side-pieces. If these are too long they will be useless; and, if they are too short, they will exert unpleasant pressure either upon the nose, or behind the ears, or in both places.

The materials of which spectacle frames should be made must depend partly upon the price, partly upon the vocations of the wearer, partly upon the general health. Steel is cheap, and moderately durable in ordinary circumstances, but it will rapidly rust, or will lose its elasticity and allow the frames to become weak and wobbly, if it is exposed to damp or to sea air. In some persons, especially those of a rheumatic constitution, the same effect is produced by the secretions of the skin, which in them are apt to be of an acid character. Of all possible materials, gold is the best when its greater cost is not prohibitory; and, whenever spectacles are worn constantly, this greater cost will soon be repaid. Moreover, gold frames have the advantage that, even if they are broken or worn out, the material will retain its value, and will partly cover the expense of providing new ones.

With regard to the material of the lenses themselves, the choice rests between glass and pebble; and on this score there is not much to add to what has been written on page 47. The higher price of pebble lenses is repaid by their greater hardness, which renders them less liable to be scratched or broken; and this character, it need hardly be said, is more important in convex than in concave lenses, because the former, by their shape, are more exposed to injury than the latter. It may be said, as a general rule, that there is no optical difference, in the present day, between the two materials ; but that pebble is decidedly advantageous for lenses which are likely to be worn continuously or for a long period, while glass is quite good enough for any which are likely soon to be exchanged for a different power. Among the other advantages of pebble, its hardness allows it to be wiped with a pocket-handkerchief ; while glass lenses should only be wiped with a bit of clean wash leather reserved for the purpose. The dust and atmospheric particles on the handkerchief will not touch the surface of the pebble, but they would soon scratch the glass beyond recovery.

More important by far than the material, either of the frame or of the lenses, is the care which should be taken of the spectacles by their owner. Spectacles which are suitable and useful, which can be worn with perfect comfort and which afford good vision, are things to be cherished and esteemed. They should have a substantial and secure case, capable of protecting them against dust or accidental breakage, and renewed as often as the infirmities of age threaten to impair its qualities. Better than a case into which they are thrust, and from which they are dragged, is one made to open with a lid, like a little flat box, and fastened with a strong snap. It should be delicately lined with velvet or with satin. As soon as the spectacles are taken off, they should be folded and put into their case, and the case into the pocket or other receptacle. For spectacles to be treated with neglect and contumely, thrown about anywhere or anyhow, left on tables, on mantelpieces, or on floors, exposed to manifold risks of being scratched or bent or broken, is a mournful sight to any who estimate them at their true value. They are essential to the proper exercise of vision by a large proportion of the inhabitants of civilised countries; and, this being so, they should be treated as carefully and as respectfully as the eyes themselves. Those who most neglect are generally the least ready to replace them; and are often to be seen struggling with type or with needlework through the obscuring media of scratched or dirty lenses, and suffering therefrom according to their deserts. A short time ago, being in an optician's shop, I could not help feeling an amused sympathy with an indignant salesman, who tendered to a lady customer a small piece of washleather, in a natty silk cover, for wiping her spectacle lenses. She was charmed at the sight of what was to her a novelty, and exclaimed to a friend who accompanied her that it was quite too nice, and that it would also do to rub and clean some of her ornaments. The shopman, fancying himself appealed to on the question, replied, with some heat, " Oh, certainly, you can rub your poker and tongs with it if you like."

It is perhaps hardly necessary to say, after the full explanation of the action of lenses which has been given in earlier portions of the volume, that one lens of a given number of dioptrics is precisely like another of the same number, supposing both to be correctly ground ; and that the spectacles which are advertised, under various ill-sounding names, as possessing special characteristics, have in reality no special characteristics whatever, and are only to be regarded as the basis of endeavours, on the part of designing persons, to prey upon the credulity of the ignorant. Spectacles with coloured lenses are especially to be avoided except under surgical advice; for, although they are useful in certain diseased conditions, they are never otherwise than injurious to perfectly healthy eyes, which are deprived by them of their natural and proper stimulant, white or solar light, and are kept in a state of artificial twilight or of unnaturally coloured illumination. Coloured spectacles are only useful, in health, to persons who are exposed to excessive glare, as in crossing a snow-covered country or in sailing under a tropical sun. In such cases, it is of the first importance that the coloured glasses should be tinted only with cobalt blue of a sufficient depth, and that they should be large enough to afford complete protection from the intrusion of lateral rays when the eyes are turned towards one side. I have found by experience, in travelling over snow in sunshine, that small blue glasses are hurtful rather than beneficial. At every sideway glance the glare of white light is admitted to the eyes; and is all the more irritating from the contrast which it offers to the sober hues seen on looking through the glasses directly to the front. The frames shown in Fig. 52, having side-pieces furnished with glasses, afford a ready means of completely protecting the eyes laterally when such protection is required. Cups of wire gauze, with or without glasses let into the front of them, and various similar contrivances which it

would be tedious to enumerate, have been devised for
the protection of the eyes of travellers and others under
various circumstances of exposure ; and of all of these
it may be said, even if they are not particularly useful,
that they are at least harmless if not worn too continuously.   In snow, of course, the object is to protect
the eyes from glare reflected up from below ; but there
are some diseased conditions, especially the early stages
of cataract, in which vision may for a time be assisted
by the dilatation of the pupil produced by diminishing
the amount of light which falls into it from above.
Whenever a patient sees better in a bright light by
shading the eyes with the hand, and by thus cutting off
the descending rays, the same assistance may be afforded
by a very useful but very ugly contrivance known as a
pair of "pents," and consisting of little black screens, of
crape   or   gauze,   projecting   forwards   from   the   upper
margins of the rings of a spectacle frame, and combined
with lenses or not, according to the other requirements
of the case.   But all appliances of this kind fall under
the head of resources for the treatment of unnatural or
diseased conditions ; and as such they lie wholly beyond
the scope and intention of the present treatise.

INDEX.

A

ACCOMMODATION, 69; diagrams of, 72, 74; decline of, 73; in relation with convergence, 85, 88; spasm of, 111
Acuteness of vision, 58; in childhood, 172; increased by cultivation, 229
Agnew, Dr., on myopia in schools, 102
Alcohol, effects of, on the sight, 227
Alum screen, 212
Ametropia, 63
Angle, visual, 56
Anterior chamber, 7
Aqueous humour, 7
Arlt, Professor, on writing, 179; on books, 181
Artificial light, characteristics of, 194; faults of, 200; Barff on, 205; Bouchardat on, 207
Artisans, use of spectacles by, 79
Asthenopia, 115, 139; accommodative or muscular, 141; Donders on, 141; case of, 143; prisms in, 145; Dr. Dyer on, 149
Astigmatism, 68, 123; correction of, 129; influence of on sight, 132, 137; test-types for, 135; spectacles for, 137
Attention, visual, education of, 175
Axis of the globe, 5; of vision, 9

B

BARFF, Professor, on artificial light, 205
Beer, Professor, on education, 180
Blind-spot, 60
Blinds, colour of, 193
Blue chimney-glasses, effects of, 200
Bouchardat, Professor, on artificial light, 207

C

CENTRE of rotation of eyeball, 6
Chamber, anterior and posterior, 7
Children, imperfect vision of, 172, 185; contagious eye diseases of, 185
Choroid, 6
Ciliary processes, 7
Cleansing the eyes of infants, 167
Clock-dials, luminous, 208
Colour-blindness, of peripheral parts of retina, 61; general, 155; Holmgren's test for, 157; temporary, 160
Colours, Newton's discovery concerning, 152; the different refrangibilities and wave-lengths of, 153; primary, 154; effects of upon the eyes, 163; of window-blinds, 193; different heating effects of, 198

LONDON:
R. CLAY, SONS, AND TAYLOR,
BREAD STREET HILL.

THE OLD LIGHTHOUSE WAS BUILT UPON A
PORTION OF THE REEF WHICH, IN ORDINARY
WEATHER, IS JUST AT THE LEVEL OF HIGH
WATER, AND WHICH AFFORDS NO MORE THAN
ROOM FOR THE STRUCTURE PLACED UPON
IT; BUT THE NEW ONE WILL STAND UPON

U L H P C D E O U P

This page may be detached from the volume and hung up against a wall in a good light as a means of testing vision. A child or other person, at a distance of seven feet, should read the words and letters fluently.

BEDFORD STREET, COVENT GARDEN, LONDON,
March, 1879.

MACMILLAN & CO.'S MEDICAL CATALOGUE.

WORKS in PHYSIOLOGY, ANATOMY, ZOOLOGY, BOTANY, CHEMISTRY, PHYSICS, MIDWIFERY, MATERIA MEDICA, and other Professional Subjects.

ALLBUTT (T. C.)—ON THE USE OF THE OPHTHALMOSCOPE in Diseases of the Nervous System and of the Kidneys ; also in certain other General Disorders. By THOMAS CLIFFORD ALLBUTT, M.A., M.D., Cantab., Physician to the Leeds General Infirmary, Lecturer on Practical Medicine, &c., &c. 8vo. 15s.

ANDERSON.—Works by DR. McCALL ANDERSON, Professor of Clinical Medicine in the University of Glasgow, and Physician to the Western Infirmary and to the Wards for Skin Diseases.

ON THE TREATMENT OF DISEASES OF THE SKIN : with an Analysis of Eleven Thousand Consecutive Cases. Crown 8vo. 5s.

LECTURES ON CLINICAL MEDICINE. With Illustrations. 8vo. 10s. 6d.

ON THE CURABILITY OF ATTACKS OF TUBERCULAR PERITONITIS AND ACUTE PHTHISIS (Galloping Consumption). Crown 8vo. 2s. 6d.

ANSTIE.—ON THE USE OF WINES IN HEALTH AND DISEASE. By F. E. ANSTIE, M.D., F.R.S., late Physician to Westminster Hospital, and Editor of *The Practitioner*. Crown 8vo. 2s.

BALFOUR.—ELASMOBRANCH FISHES ; a Monograph on the Development of. By F. M. BALFOUR, M.A., Fellow and Lecturer of Trinity College, Cambridge. / With Plates. 8vo. 21s.

BARWELL.—ON CURVATURES OF THE SPINE : their Causes and treatment. By RICHARD BARWELL, F.R.C.S., Surgeon and late Lecturer on Anatomy at the Charing Cross Hospital. Third Edition, with additional Illustrations. Crown 8vo. 5s.

BASTIAN.—Works by H. CHARLTON BASTIAN, M.D., F.R.S., Professor of Pathological Anatomy in University College, London, &c. :—

THE BEGINNINGS OF LIFE : Being some Account of the Nature, Modes of Origin, and Transformations of Lower Organisms. In Two Volumes. With upwards of 100 Illustrations. Crown 8vo. 28s.

EVOLUTION AND THE ORIGIN OF LIFE. Crown 8vo. 6s. 6d.

ON PARALYSIS FROM BRAIN DISEASE IN ITS COMMON FORMS. Illustrated. Crown 8vo. 10s. 6d.

" It would be a good thing if all such lectures were as clear, as systematic, and as interesting. . . . . It is of interest not only to students but to all who make nervous diseases a study."—*Journal of Mental Science.*

BUCKNILL.—HABITUAL DRUNKENNESS AND INSANE DRUNKARDS. By J. C. BUCKNILL, M.D. Lond., F.R.S., F.R.C.P., late Lord Chancellor's Visitor of Lunatics. Crown 8vo. 2s. 6d.

CARTER.—Works by R. BRUDENELL CARTER, F.R.C.S., Ophthalmic Surgeon to St. George's Hospital, &c.

A PRACTICAL TREATISE ON DISEASES OF THE EYE. With Illustrations. 8vo. 16s.

" No one will read Mr. Carter's book without having both his special and general knowledge increased."—*Lancet.*

ON DEFECTS OF VISION WHICH ARE REMEDIABLE BY OPTICAL APPLIANCES. Lectures at the Royal College of Surgeons. With numerous Illustrations. 8vo. 6s.

CHRISTIE.—CHOLERA EPIDEMICS IN EAST AFRICA. An Account of the several Diffusions of the Disease in that country from 1821 till 1872, with an Outline of the Geography, Ethnology, and Trade Connections of the Regions through which the Epidemics passed. By J. Christie, M.D., late Physician to H.H. the Sultan of Zanzibar. With Maps. 8vo. 15s.

COOKE (JOSIAH P., Jun.).—FIRST PRINCIPLES OF CHEMICAL PHILOSOPHY. By Josiah P. Cooke, Jun., Ervine Professor of Chemistry and Mineralogy in Harvard College. Third Edition, revised and corrected. Crown 8vo. 12s.

CREIGHTON. — CONTRIBUTIONS TO THE PHYSIOLOGY AND PATHOLOGY OF THE BREAST AND ITS LYMPHATIC GLANDS. By Charles Creighton, M.D., Demonstrator of Anatomy in the University of Cambridge. With Illustrations. 8vo. 9s.

"It is impossible not to see at once that the work is deserving of all praise, both from the originality and from the care which has been bestowed upon it."—*Practitioner.*

FLOWER (W. H.).—AN INTRODUCTION TO THE OSTEOLOGY OF THE MAMMALIA. Being the substance of the Course of Lectures delivered at the Royal College of Surgeons of England in 1870. By W. H. Flower, F.R.S., F.R.C.S., Hunterian Professor of Comparative Anatomy and Physiology. With numerous Illustrations. Second Edition, revised and enlarged. Crown 8vo. 10s. 6d.

FOSTER.—Works by Michael Foster, M.D., F.R.S. :—

A TEXT BOOK OF PHYSIOLOGY, for the use of Medical Students and others. Second Edition, revised and enlarged, with additional Plates and Illustrations. 8vo. 21s.

"Dr. Foster has combined in this work the conflicting desiderata in all textbooks—comprehensiveness, brevity, and clearness. After a careful perusal of the whole work we can confidently recommend it, both to the student and the practitioner as being one of the best text-books on physiology extant."—*Lancet.*

A PRIMER OF PHYSIOLOGY. Illustrated. 18mo. 1s.

FOSTER and LANGLEY.—AN ELEMENTARY COURSE OF PRACTICAL PHYSIOLOGY. By Michael Foster, M.D., F.R.S., assisted by J. N. Langley, B.A. Third Edition, enlarged. Crown 8vo. 6s.

"Equipped with a text-book such as this . . . . the beginner cannot fail to acquire a real, though of course elementary, knowledge of the leading facts and principles of Physiology."—*Academy.*

FOSTER and BALFOUR.—ELEMENTS OF EMBRYOLOGY. By Michael Foster, M.D., F.R.S., and F. M. Balfour, M.A., Fellow of Trinity College, Cambridge. With numerous Illustrations. Part I. Crown 8vo. 7s. 6d.

"Both text and illustrations are alike remarkable for their clearness and freedom from error, indicating the immense amount of labour and care expended in the production of this most valuable addition to scientific literature."—*Medical Press and Circular.*

FOTHERGILL. — Works by J. Milner Fothergill, M.D., M.R C.P., Assistant Physician to the Victoria Park Chest Hospital, and to the West London Hospital :—

THE PRACTITIONER'S HANDBOOK OF TREATMENT : or, THE PRINCIPLES OF RATIONAL THERAPEUTICS. 8vo. 14s.

"We have every reason to thank the author for a practical and suggestive work." —*Lancet.*

THE ANTAGONISM OF THERAPEUTIC AGENTS, AND WHAT IT TEACHES. The Essay to which was awarded the Fothergillian Gold Medal of the Medical Society of London for 1878. Crown 8vo. 6s.

FOX.—Works by WILSON FOX, M.D., Lond., F.R.C.P., F.R.S., Holme Professor of Clinical Medicine, University College, London, Physician Extraordinary to her Majesty the Queen, &c. :—

DISEASES OF THE STOMACH : being a new and revised Edition of " THE DIAGNOSIS AND TREATMENT OF THE VARIETIES OF DYSPEPSIA." 8vo. 8s. 6d.

ON THE ARTIFICIAL PRODUCTION OF TUBERCLE IN THE LOWER ANIMALS. With Coloured Plates. 4to. 5s. 6d.

ON THE TREATMENT OF HYPERPYREXIA, as Illustrated in Acute Articular Rheumatism by means of the External Application of Cold. 8vo. 2s. 6d.

GALTON (D.).—AN ADDRESS ON THE GENERAL PRINCIPLES WHICH SHOULD BE OBSERVED IN THE CONSTRUCTION OF HOSPITALS. By DOUGLAS GALTON, C.B., F.R.S. Crown 8vo. 3s. 6d.

GEGENBAUR.—ELEMENTS OF COMPARATIVE ANATOMY. By CARL GEGENBAUR, Professor of Anatomy and Director of the Anatomical Institute, Heidelberg. A translation by F. JEFFREY BELL, B.A., revised, with Preface by E. RAY LANKESTER, M.A., F.R.S., Professor of Zoology and Comparative Anatomy in University College, London. With numerous Illustrations. Medium 8vo. 21s.

GRIFFITHS.—LESSONS ON PRESCRIPTIONS AND THE ART OF PRESCRIBING. By W. HANSEL GRIFFITHS, Ph.D., L.R.C.P.E. New Edition. 18mo. 3s. 6d.

" We recommend it to all students and junior members of the profession who desire to understand the art of prescribing."—*Medical Press.*

HANBURY.—SCIENCE PAPERS, chiefly Pharmacological and Botanical. By DANIEL HANBURY, F.R.S. Edited with Memoir by JOSEPH INCE, F.L.S., F.C.S. 8vo. 14s.

HOOD (Wharton.).— ON BONE-SETTING (so-called), and its Relation to the Treatment of Joints Crippled by Injury, Rheumatism, Inflammation, &c., &c. By WHARTON P. HOOD, M.D., M.R.C.S. Crown 8vo. Illustrated. 4s. 6d.

"Dr. Hood's book is full of instruction, and should be read by all surgeons."— *Medical Times.*

HOOKER (Dr.).—THE STUDENT'S FLORA OF THE BRITISH ISLANDS. By Sir J. D. HOOKER, K.C.S.I., C.B., M.D., D.C.L., President of the Royal Society. Second Edition, revised and corrected. Globe 8vo. 10s. 6d.

HUMPHRY.—Works by G. M. HUMPHRY, M.D., F.R.S., Professor of Anatomy in the University of Cambridge, and Honorary Fellow of Downing College :—

THE HUMAN SKELETON (including the Joints). With 260 Illustrations drawn from Nature. Medium 8vo. 28s.

OBSERVATIONS IN MYOLOGY. Illustrated. 8vo. 6s.

THE HUMAN FOOT AND HAND. Illustrated. Fcap. 8vo. 4s. 6d.

HUXLEY and MARTIN. — A COURSE OF PRACTICAL INSTRUCTION IN ELEMENTARY BIOLOGY. By T. H. HUXLEY, LL.D. Sec. R.S., assisted by H. N. MARTIN, M.B., D.Sc. New Edition, revised. Crown 8vo. 6s.

" To intending medical students this book will prove of great value."—*Lancet.*

HUXLEY (Professor).—LESSONS IN ELEMENTARY PHYSIOLOGY. By T. H. HUXLEY, LL.D., F.R.S. With numerous Illustrations. New Edition. Fcap. 8vo. 4s. 6d.

KEETLEY.—THE STUDENT'S GUIDE TO THE MEDICAL PROFESSION. By C. B. KEETLEY, F.R.C S., Assistant Surgeon to the West London Hospital. With a Chapter for Women Students. By Mrs. GARRETT ANDERSON. Crown 8vo. 2s. 6d.

KÜHNE.—ON THE PHOTOCHEMISTRY OF THE RETINA AND ON VISUAL PURPLE. Translated from the German of Dr. KÜHNE, and Edited, with Notes, by MICHAEL FOSTER, M.D., F.R.S. 8vo. 3s. 6d.

LANKESTER.—COMPARATIVE LONGEVITY IN MAN AND THE LOWER ANIMALS. By E. RAY LANKESTER B.A. Crown 8vo. 4s. 6d.

LEISHMAN.—A SYSTEM OF MIDWIFERY, including the Diseases of Pregnancy and the Puerperal State. By WILLIAM LEISHMAN. M.D., Regius Professor of Midwifery in the University of Glasgow : Physician to the University Lying-in Hospital : Fellow and late Vice-President of the Obstetrical Society of London, &c., &c. 8vo. Illustrated. Second and Cheaper Edition. 21s.

MACLAGAN. — THE GERM THEORY APPLIED TO THE EXPLANATION OF THE PHENOMENA OF DISEASE. By T. MACLAGAN, M.D. 8vo. 10s. 6d.
"We think it well that such a book as this should be written. It places before the reader in clear and unmistakable language what is meant by the germ theory of disease."—*Lancet.*

MACNAMARA.—Works by C. MACNAMARA, F.C.U., Surgeon to Westminster Hospital :—

A HISTORY OF ASIATIC CHOLERA. Crown 8vo. 10s. 6d.
" A very valuable contribution to medical literature, and well worthy of the place which it is sure to assume as the standard work on the subject."—*Medical Examiner.*

DISEASES OF BONE.—CLINICAL LECTURES. Crown 8vo. 5s.

MACPHERSON.—Works by JOHN MACPHERSON, M.D. :—
THE BATHS AND WELLS OF EUROPE : their Action and Uses. With Notices of Climatic Resorts and Diet Cures. With a Map. New Edition, revised and enlarged. Extra fcap. 8vo. 6s. 6d.

OUR BATHS AND WELLS : The Mineral Waters of the British Islands. With a List of Sea-Bathing Places. Extra fcap. 8vo. 3s. 6d.

MANSFIELD (C. B.).—A THEORY OF SALTS. A Treatise on the Constitution of Bipolar (two-membered) Chemical Compounds. By the late CHARLES BLACHFORD MANSFIELD. Crown 8vo. 14s.

MAUDSLEY.—Works by HENRY MAUDSLEY, M.D., Professor of Medical Jurisprudence in University College, London :—

BODY AND MIND : An Inquiry into their Connection and Mutual Influence, specially in reference to Mental Disorders : being the Gulstonian Lectures for 1870. Delivered before the Royal College of Physicians. New Edition, with Psychological Essays added. Crown 8vo. 6s. 6d.

THE PHYSIOLOGY OF MIND. Being the First Part of a Third Edition, revised, enlarged, and in great part re-written, of " The Physiology and Pathology of Mind." Crown 8vo. 10s. 6d.

THE PATHOLOGY OF MIND. [*In the Press.*]

MIALL.—STUDIES IN COMPARATIVE ANATOMY.
No. I.—The Skull of the Crocodile. By L. C. MIALL, Professor of Biology in the Yorkshire College of Science. 8vo. 2s. 6d.
No. II.—The Anatomy of the Indian Elephant. By L. C. MIALL and F. GREENWOOD, Curator of the Leeds School of Medicine. Illustrated. 8vo 5s

MIVART (St. George).—Works by ST. GEORGE MIVART, F.R.S., &c., Lecturer in Comparative Anatomy at St. Mary's Hospital :—

ON THE GENESIS OF SPECIES. Second Edition, to which notes have been added in reference and reply to Darwin's "Descent of Man." With numerous Illustrations. Crown 8vo. 9s.

LESSONS IN ELEMENTARY ANATOMY. With upwards of 400 Illustrations. New Edition. Fcap. 8vo. 6s. 6d.

"It may be questioned whether any other work on anatomy contains in like compass so proportionately great a mass of information."—Lancet.

M'KENDRICK.—OUTLINES OF PHYSIOLOGY IN ITS RELA-TIONS TO MAN. By JOHN GRAY M'KENDRICK, M.D., F.R.S.E., Professor of the Institute of Medicine and Physiology in the University of Glasgow. Illustrated. Crown 8vo. 12s. 6d.

MUIR.—PRACTICAL CHEMISTRY FOR MEDICAL STUDENTS. Specially arranged for the first M. B. Course. By M. M. PATTISON MUIR, F.R.S.E., Prælcetor in Chemistry, Caius College, Cambridge. Fcap. 8vo. 1s. 6d.

"This little book will aid the student not only to pass his professional examination in practical Chemistry more easily, but will give him such an insight into the subject as will enable him readily to extend his knowledge of it should time and inclination permit."—Practitioner.

OLIVER.—LESSONS IN ELEMENTARY BOTANY. By DANIEL OLIVER, F.R.S., F.L.S., Professor of Botany in University College, London, and Keeper of the Herbarium and Library of the Royal Gardens, Kew. With nearly 200 Illustrations. New Edition. Fcap. 8vo. 4s. 6d.

PARKER and BETTANY.—THE MORPHOLOGY OF THE SKULL. By W. K. PARKER, F.R.S., Hunterian Professor, Royal College of Surgeons, and G. T. BETTANY, M.A., B.Sc., Lecturer on Botany in Guy's Hospital Medical School. Crown 8vo. 10s. 6d.

PETTIGREW.—THE PHYSIOLOGY OF THE CIRCULATION IN PLANTS, IN THE LOWER ANIMALS, AND IN MAN. By J. BELL PETTIGREW, M.D.. F.R.S., etc. Illustrated by 150 Woodcuts. 8vo. 12s.

"A more original, interesting, exhaustive, or comprehensive treatise on the circulation and the circulatory apparatus in plants, animals, and man, has never, we are certain, been offered for the acceptance of the anatomist physiologist or student of medicine."—Veterinary Journal.

PIFFARD.—AN ELEMENTARY TREATISE ON DISEASES OF THE SKIN, for the Use of Students and Practitioners. By H. G. PIFFARD, M.D., Professor of Dermatology in the University of the City of New York, &c. With Illustrations. 8vo. 16s.

RADCLIFFE.—Works by CHARLES BLAND RADCLIFFE, M.D., F.R.C.P., Physician to the Westminster Hospital, and to the National Hospital for the Paralysed and Epileptic :—

VITAL MOTION AS A MODE OF PHYSICAL MOTION. Crown 8vo. 8s. 6d.

PROTEUS : OR UNITY IN NATURE. Second Edition. 8vo. 7s. 6d.

RANSOME.—ON STETHOMETRY. Chest Examination by a more Exact Method with its Results. With an Appendix on the Chemical and Microscopical Examination of Respired Air. By ARTHUR RANSOME, M.D. With Illustrations. 8vo. 10s. 6d.

"We can recommend his book not only to those who are interested in the graphic method, but to all who are specially concerned in the treatment of diseases of the chest."—British Medical Journal.

REYNOLDS (J. R.).—A SYSTEM OF MEDICINE. Edited by J. RUSSELL REYNOLDS, M.D., F.R.S. London. In 5 Vols. Vols. I. to III., 25s. each ; Vol IV., 21s. ; Vol. V., 25s.

REYNOLDS (J. R.).—*continued.*

Vol. I.—Part I. General Diseases, or Affections of the Whole System. Part II. Local Diseases, or Affections of Particular Systems. § I.—Diseases of the Skin.

Vol. II.—Part II. Local Diseases (continued). § I.—Diseases of the Nervous System. § II.—Diseases of the Digestive System.

Vol. III.—Part II. Local Diseases (continued). § II.—Diseases of the Digestive System (continued). § III.—Diseases of the Respiratory System.

Vol. IV.—Diseases of the Heart. Part II. Local Diseases (continued). § IV.—Diseases of the Organs of Circulation.

Vol. V.—Diseases of the Organs of Circulation.—Diseases of the Vessels.—Diseases of the Blood-Glandular System.—Diseases of the Urinary Organs.—Diseases of the Female Reproductive Organs.—Diseases of the Cutaneous System.

Also, now publishing in MONTHLY PARTS, Price 5s. each, to be completed in 24 Parts. (Part 1, April 1st, 1879.)

RICHARDSON.—Works by B. W. RICHARDSON, M.D., F.R.S. :—

DISEASES OF MODERN LIFE. Fifth and Cheaper Edition. Crown 8vo. 6s.

ON ALCOHOL. New Edition. Crown 8vo. 1s.

HYGEIA, A CITY OF HEALTH. Crown 8vo. 1s.

THE FUTURE OF SANITARY SCIENCE. Crown 8vo. 1s.

TOTAL ABSTINENCE. A course of addresses. Crown 8vo. 3s. 6d.

ROSCOE.—Works by HENRY ROSCOE, F.R.S., Professor of Chemistry in Owens College, Manchester :—

LESSONS IN ELEMENTARY CHEMISTRY, INORGANIC AND ORGANIC. With numerous Illustrations, and Chromolithographs of the Solar Spectrum and of the Alkalies and Alkaline Earths. New Edition. Fcap. 8vo. 4s. 6d.

CHEMICAL PROBLEMS, adapted to the above. By Professor T. E. THORPE, M.D., F.R.S.E., with Preface by Professor Roscoe. Fifth Edition, with Key. 18mo. 2s.

PRIMER OF CHEMISTRY. Illustrated. 18mo. 1s.

ROSCOE and SCHORLEMMER.—A TREATISE ON CHEMISTRY. By Professors ROSCOE and SCHORLEMMER. Vol. I. The Non-Metallic Elements. With Numerous Illustrations and Portrait of Dalton. 8vo. 21s. Vol. II. Metals. Part I. With numerous Illustrations. 8vo. 21s.

SCHORLEMMER.—A MANUAL OF THE CHEMISTRY OF THE CARBON COMPOUNDS, OR ORGANIC CHEMISTRY. By C. SCHORLEMMER, F.R.S., Lecturer in Organic Chemistry in Owens College, Manchester. 8vo. 14s.

SEATON.—A HANDBOOK OF VACCINATION. By EDWARD C. SEATON, M.D., Medical Inspector to the Privy Council. Extra fcap. 8vo. 8s. 6d.

SEILER.—MICRO-PHOTOGRAPHS IN HISTOLOGY, Normal and Pathological. By CARL SEILER, M.D., in conjunction with J. GIBBONS HUNT, M.D., and J. G. RICHARDSON, M.D. 4to. 31s. 6d.

SPENDER.—THERAPEUTIC MEANS FOR THE RELIEF OF PAIN. Being the Prize Essay for which the Medical Society of London awarded the Fothergillian Gold Medal in 1874. By JOHN KENT SPENDER, M.D. Lond., Surgeon to the Mineral Water Hospital, Bath. 8vo. 8s. 6d.

STEWART (B.).—LESSONS IN ELEMENTARY PHYSICS. By
BALFOUR STEWART, F.R.S., Professor of Natural Philosophy in Owens
College, Manchester. With Numerous Illustrations and Chromolithograph of
the Spectra of the Sun, Stars, and Nebulæ. New Edition. Fcap. 8vo. 4s. 6d.

PRIMER OF PHYSICS. By the same Author. Illustrated. 18mo. 1s.

TUKE.—INSANITY IN ANCIENT AND MODERN LIFE, with
Chapters on its Prevention. By D. HACK TUKE, M.D., F.R.C.P. Crown 8vo.
6s.

"This work exhibits deep research in various directions, and teems with allusions
and quotations which prove the author to be not only an accomplished psychological physician, but a scholar of no mean order."—*Medical Times.*

WEST.—HOSPITAL ORGANISATION. With special reference to
the organisation of Hospitals for Children. By CHARLES WEST, M.D. Founder
of, and for twenty-three years Physician to, the Hospital for Sick Children.
Crown 8vo. 2s. 6d.

WURTZ.—A HISTORY OF CHEMICAL THEORY from the Age
of Lavoisier down to the present time. By AD. WURTZ. Translated by HENRY
WATTS, F.R.S. Crown 8vo. 6s.

MANUALS FOR STUDENTS.

THE MORPHOLOGY OF THE SKULL. By W. K. PARKER,
F.R.S., Hunterian Professor, Royal College of Surgeons, and G. T. BETTANY,
B.Sc., Lecturer on Botany in Guy's Hospital Medical School. Illustrated.
Crown 8vo. 10s. 6d.

THE OSTEOLOGY OF THE MAMMALIA: A Series of
Lectures by Prof. W. H. FLOWER, F.R.S., F.R.C.S. With numerous Illustrations. New Edition, enlarged. Crown 8vo. 10s. 6d.

THE ELEMENTS OF EMBRYOLOGY. By MICHAEL FOSTER,
M.D., F.R.S., and F. M. BALFOUR, M.A. Part I. 7s. 6d.

PRACTICAL PHYSIOLOGY: an Elementary Course of. By Dr.
M. FOSTER, assisted by J. LANGLEY. New Edition. Crown 8vo. 6s.

ELEMENTARY BIOLOGY: a Course of Practical Instruction
in. By Prof. HUXLEY and H. N. MARTIN. New Edition. Crown 8vo. 6s.

PHYSIOGRAPHY: an Introduction to the Study of Nature. By
Prof. HUXLEY, F.R.S. With Coloured Plates and Woodcuts. New Edition.
Crown 8vo. 7s. 6d.